A Ticket to Hell:
ON OTHER MEN'S SINS

A Ticket to Hell:
ON OTHER MEN'S SINS

The Fourth Jon and Teresa Zachery Story

J. J. ZERR

PRIMIX
PUBLISHING
THE WRITE CHOICE

Primix Publishing
11620 Wilshire Blvd
Suite 900, West Wilshire Center, Los Angeles, CA, 90025
www.primixpublishing.com
Phone: 1 (888) 585-7476

© 2021 J. J. Zerr. All rights reserved.

No part of this book may be reproduced, stored in a retrieval system, or transmitted by any means without the written permission of the author.

Published by Primix Publishing   02/16/2021

ISBN: 978-1-953397-97-3(sc)
ISBN: 978-1-953397-98-0(hc)
ISBN: 978-1-953397-99-7(e)

Any people depicted in stock imagery provided by iStock are models, and such images are being used for illustrative purposes only.

Certain stock imagery © iStock.

Because of the dynamic nature of the Internet, any web addresses or links contained in this book may have changed since publication and may no longer be valid. The views expressed in this work are solely those of the author and do not necessarily reflect the views of the publisher, and the publisher hereby disclaims any responsibility for them.

Also by J. J. Zerr

The Jon Zachery stories:

The Ensign Locker

Sundown Town Duty Station

The Junior Officer Bunkroom

Other novels:

Noble Deeds

The Happy Life of Preston Katt

Guerilla Bride

The Ghosts of Chateau du Chasse

Short story collection:

War Stories

God bless editors and my Coffee and Critique Bubbas and Bubbettes.

For the real-world Warhorse spouses. Real-world war heroes. Thank you for your service.

Author's Note

A cast of principal characters, a listing of naval ranks, and a list of US Navy acronyms and terms, as used in this story, are included in the end matter.

PART 1
ON OTHER MEN'S SINS

Chapter 1

Teresa Zachery checked the bathroom mirror for puffy cheeks. Not yet. Three months pregnant. She still had a month or so, or maybe a week.

Her husband, Jon, often told her, "You remind me of Michelangelo's Mary in his *Pieta*."

Pieta Mary was a beautiful woman, whose cheeks were not puffed out. It was nice being married to a man who considered her beautiful.

Teresa sighed. *Puffy cheeks.* The subject could not be considered without an erumpent memory of growing up. Mother had harped on her about her weight throughout grade school and high school. Now she studied her profile in a full-length mirror and compared herself, unfavorably, to her neighbor—beautiful, blue-eyed, blonde, shapely Amy Allison.

Teresa would never consider herself beautiful. She wasn't blonde. Her hair and eyes were brown, evidence she couldn't be as beautiful as her neighbor. Although she and Amy wore the same dress size, that was irrelevant and could not override what had been implanted in her head: "You're fat." Mother never said it directly, but there were ways to imply things so strongly that the implication drove the listener to an inescapable conclusion.

Early in their marriage, she told Jon she was fat and that she'd been fat her whole life.

"You certainly weren't fat in high school," Jon said. "You sure as shooting aren't overweight now. Where did you get this idea?"

Her husband was not a talker. But give him a puzzle to work, and he gnawed at it until the solution appeared. She admitted the idea had come from her mother. Jon had retrieved an album with pictures of Teresa growing up. He showed her a number of them.

"What I see in these is a girl who is not fat and who is not happy. That last part hurts my heart, Teresa Velmer Zachery."

Teresa Velmer Zachery looked into the brown eyes of her image. *You are thirty years old, married, have two children, are pregnant, and you still have mommy issues.*

The calendar. Today was a big day.

She left the bathroom and entered the kitchen.

Her calendar was pinned to a corkboard on the wall between the washer and dryer and the door to the carport. First thing in the morning, after the bathroom, she checked it for scheduled activities and to X off another day. Her husband, US Navy Lieutenant Jon Zachery, and her neighbor Mike Allison served aboard the aircraft carrier USS *Solomons* on the other side of the world, in the Tonkin Gulf, flying combat missions over Vietnam. Tomorrow, when she X-ed off today, the circled date, it would mark seven months since the deployment began.

The circled date marked the last day of combat flying. It would still take a month before the carrier crossed the Pacific and returned the crew to their families.

"The Pacific," Jon liked to say, "is the biggest puddle of water on earth. Takes a while to cross it."

Another long month to wait, but this one would be without the anguish over him flying combat missions sitting on her heart like a chunk of lead.

She rested her hands on her tummy. Three months pregnant. She'd spent a week with Jon in Hong Kong during the USS *Solomons*'s port visit, well, three months prior.

Daniel. Almost two years ago, she'd gone into labor at seven months. Her son's little lungs had not been sufficiently developed, and she—they—lost him.

Now her doctor wanted her to avoid stress, but she worried she wouldn't be able to carry this baby to term. That carried a baby buggy full of stress—as did her husband being a navy carrier pilot flying combat missions over Vietnam. Daughter Jennifer, at five, was bright, inquisitive, and happy. Teresa wanted to do the right things for her, to protect her and enable her to develop her gifts. Why would there be stress in that? Three-year-old Edgar Jon was a very different child—needier in many ways. Was she doing the right things for him? No stress there.

Jon had tagged the new baby with "Little Pootzer." With the others, they'd picked out a boy's and a girl's name as soon as she knew she was pregnant. They'd never given a nickname to their other in utero children. Avoiding picking a real name for this one was perhaps his way of dealing with the loss of Daniel. Maybe he thought picking a real name so early in the pregnancy jinxed the child.

Just like circling the date on the calendar jinxed the guys on the *Solomons*. Some of the wives of the Warhorse pilots believed that and would never circle a date as she had done. Some of the pilots also believed that verbalizing the anticipation of a favorable outcome was the surest way to disappointment, by jinxing it. She was sure Jon considered jinxes to be groundless superstition; still, when he heard someone say something jinx worthy, he tapped his head three times and said, "Best wood around."

Navy aircraft carriers routinely had deployments lengthened, port visits canceled, and flying combat missions extended. Even in peacetime, those things happened. In wartime, schedules were less than guesses.

I shouldn't have circled the date.

But she'd needed something to look forward to, to mark the end of the most awful part of him being gone. For her, the circled date was not filled with jinx; it was filled with hope.

"Father God, who art in heaven," she whispered. "My faith is in You. Please watch over Jon extra closely today and bring him home on time. Please and amen."

A calm settled over her as if she'd stepped from a goose bump–filled, cool shadow into the sunlight. Jon was and would be okay. She just knew.

"Thank you, Lord," she whispered to the circled date.

She patted Little Pootzer.

Daddy's okay.

Walking toward her bedroom, she stopped at the children's and checked on them. Sleeping angels. Jennifer's brown hair looked like it had just been brushed.

"God created you, Teresa Velmer Zachery, and our daughter with an immunity to bedhead," Jon said once.

"You jealous?" she said.

"You betcha. It's why I keep my hair short, and EJ's too."

So many things wound up acronymized in the navy, even their son's name.

Edgar Jon, still in the terrible threes, also looked like an angel as he slept, a blond one. An awake Edgar Jon, however, quite often behaved like another kind of spiritual being.

The toys were all in the toybox. The children's room was shipshape. As was her house. Rather, it was the US Navy's house on Naval Air Station Lemoore, California, but she cared for it as if it belonged to Jon and her. If she let the care of her house slip, it was the same as letting care for her children and herself slip into slovenliness.

In her bedroom, as she undressed and then dressed for bed, she thought of Jon on the other side of the world. She hoped his last combat hop would be in daylight. Night hops scared the *bejeebers*—the pilots had their own crude language—out of carrier aviators. Teresa heard them talk at parties; with a few drinks in them, pilots would tell of the terror they felt on a catapult at night as they waited for the jolt that would hurl them into blackness.

She remembered Skunk, the story about how he got his call sign. On his first night catapult shot, he'd peed in his flight suit. He'd launched, climbed, turned to fly behind the carrier, and landed back aboard. According to him, he'd been so scared on the cat shot he had no fear left over for the landing. After his landing, the carrier ended flight ops for that day. Skunk went down to the ready room, and another pilot smelled urine. Skunks do not notice their own stink. Also, in his black hair above his left ear, there was a white apostrophe. He became Skunk.

Please, God, let Jon's last hop not *be a nighter.*

Tomorrow morning, when she X-ed off today, such a burden, so much stress would be lifted. Jon, however, wouldn't see it that way. The carrier *Solomons* and the squadrons embarked on her were all slated to decommission after the ship returned. Jon expected some no-account, dead-end job to occupy his last two years of obligation. He would view it as the navy casting him onto a trash heap, his service a thing of no value. It would puncture a hole in his soul.

Why, God, can't it be the same for both of us?

Teresa sighed as "the night of the dog poop," as Jon called it, slipped into her mind.

Late fall, 1966. Jon had just returned from a deployment to the Tonkin Gulf aboard a destroyer. Before that night, their plan was this: Jon would serve out his obligated time; he would leave the service; he would get a job as an electrical engineer; and they would live happily ever after. That night, the night of the dog poop shattered their dream.

A week after his ship returned to San Diego, Jon, nine-month-old Jennifer, and Teresa had visited her uncle Theodore and aunt Penelope Prescott north of Los Angeles. Their daughter, Christine, was away at college and not expected home. But she'd caught an opportune ride and entered her house to find the Zacherys at dinner with her parents.

"Baby killer," she snarled at Jon and ran back out. Uncle Theodore ran after her, dragged her inside again, and made her apologize to Jon. Late that night, friends of Christine trashed the Zachery car with dog poop and a garden hose.

Before that incident, Jon walked away from the TV when antiwar protest news came on. "It's not antiwar, it's anti-America," he groused. But it was on TV, where things that happened to other people were reported. The dog poop made it personal. He disagreed with the protest, and he felt obligated to demonstrate his position. It was his duty to do so. On his destroyer, he wasn't doing as much as the aviators who carried the war to the enemy. He wanted to apply for aviation, to be a carrier pilot, to do something meaningful.

When he'd said that, it drove an icicle into her chest. Carrier pilots were often killed during routine operations. When the North Vietnamese fired missiles and bullets at them, more were killed. He asked for her acceptance of his call to duty. It pushed *happily ever after* from just ahead almost all the way to unobtainable, and it caused her physical and mental pain to say, "You have to do what you think is right."

When the squadron decommissioned, Jon would see his noble call to duty, his call to do what he thought was right, as having been a waste.

Please, God, help me find the way to help Jon.

She climbed into bed and picked up her rosary from the bedside table. Before starting on the beads, she realized she had done nothing but ask the Lord for things. Teresa ticked off blessings. With her mind aimed at the task, she found a number of them and thanked Him for each one. The last: *And Jon is okay.*

She exhaled worries and sank into peace-filled darkness, her rosary beads resting just above Little Pootzer.

Chapter 2

When he donned his flight suit, Lieutenant Jon Zachery—as did all the airwing pilots—sluffed off his baptismal certificate name and US Navy rank and took on his call sign, Stretch.

Stretch and aircraft 510 slammed onto the deck of the USS *Solomons*. Throttle to full power. If the tailhook failed to snag an arresting wire, 510 would run off the end of the flight deck in two seconds. If the power was not at full, Stretch and 510 would crash into the Tonkin Gulf. But they, Stretch and 510, caught a wire. The force of the arrested landing flung him against the shoulder straps. Off to his right, a flight deck taxi director signaled: Power back, power back. Tailhook up, tailhook up. Taxi clear of the landing area, taxi clear.

Before he could comply with the third directive, the carrier started a turn to starboard. Aircraft 510 was the last to land from the last mission of the '70–'71 deployment of the USS *Solomons* to the Vietnam War. The big ship eased into the maneuver so as not to tilt the deck while airplanes were still taxiing.

Heck of a dadburned last mission!

Stretch!

He drove his mind back to the task at hand. Follow his taxi directors' signals until he was parked in his assigned spot, where twelve tiedown chains would secure 510 to the deck, he'd shut the engine down, and he'd safe his ejection seat. Then he could let his mind think of other things.

In his spot and tied down, the plane captain signaled for him to shut the engine down. Stretch held up one finger, meaning *one minute*. He ran the BIT (built-in test) function for the threat warning system. The system warned a pilot that a radar from an enemy SAM, AAA, or a MiG fighter had locked onto his plane. The BIT check indicated the system functioned properly.

Except it hadn't. During the mission, SAMs had been fired at Stretch and his flight lead, RT. Neither of them had been alerted by the warning system. Fortunately, Stretch saw the missiles, and they evaded them. SAMs were also fired at two other Warhorse pilots, and AB and Skunk had been shot down.

After six months of flying mostly low-pucker factor missions over South Vietnam and Laos, the last one spent all the adrenaline Stretch had saved up through half a year. The last launch of the cruise was a photo reconnaissance flight over southern North Vietnam. Usually, the North Viets didn't shoot at photo planes or their escorts. That day, though, they fired a bunch of SAMs and a lot of AAA. And through it all, the threat warning system issued not one peep.

Stretch shut down, climbed out of the cockpit, went straight to the bottom rear of his aircraft, and opened the door giving him access to the threat warning electronic box. Since the BIT had been good, the problem had to be with the antennas or the cables running from the box to the antennas. He unscrewed the antenna cable plug from the box and immediately saw the problem. Some of the pins in the head of the plug had been pulled out. The electronic box was not connected to the antennas. That was why the warning system had failed to warn.

Stretch hustled to RT's plane—RT had been his flight lead—and checked his warning system cable. The same discrepancy.

Judas Priest!

Only one possible explanation—sabotage!

Whoever disabled the system knew it well. One of the avionics technicians? Stretch could not imagine any one of them doing such a—

Amos Kane. He was smart enough to pull it off.

Lieutenant (junior grade) Kane, a former pilot and now a maintenance officer. The word was Kane had been a good stick and throttle jockey, but he had a rotten attitude and had mouthed off to the wrong people, and the navy jerked his wings. Now in the JOB (junior officer bunkroom), his attitude and his mouth spouting antiwar balderdash had created an atmosphere of hostility just short of violence. The five Warhorse pilots, including Stretch, who lived with him wished Amos lived someplace else.

Last mission of the deployment for the carrier. Last chance for Kane to make a statement on behalf of the antiwarriors.

Stretch ran to the island, entered, and thundered down one ladder (navy for stairs) and entered the

Intelligence Center. RT, Lieutenant Commander Robert T. Fischer, sat across a table from a young JG (lieutenant—junior grade) named Miller. The JG took notes as RT described their encounter with SAMs.

"Miller, beat it," Stretch said. "I need to talk to RT."

Miller's fair complexion sprouted red cheeks. "This is important stuff. We have to get a report off ASAP."

Stretch leaned across the table. "Get the hell outa here, or I'll throw you out."

The kid stood up, knocking over his chair. As he left the debriefing cubicle, he muttered, "Asshole pilot."

Stretch explained what he found and whom he suspected.

RT said, "Son of a bitch."

Stretch had never heard the man swear before.

RT shook his head, then looked at Stretch. "It could have been Amos. It also could have been one my techs." RT was the squadron maintenance officer, and avionics technicians worked for him. He rubbed his chin. "It could also be just about anyone who works on the flight deck. Remember the rash of dud bombs we dropped last month? We never figured out who screwed with the bomb-arming system."

Stretch knew a couple of the guys in the JOB were sure it had been Amos goddamn Kane.

"Okay, Stretch. This is what we've gotta do. First, interview Amos and all our avionics techs and flight deck personnel. Second, inspect the rest of our planes. Third, we need the other squadrons to check their planes. We need to move on this. Chop-chop. Get your cassette recorder. I'd like to record as many of the interviews as we can."

Stretch used his recorder to make voice messages for Teresa and the children, but with all the people they were going to talk to, it would be a handy tool. He descended to the hangar bay, and as he hustled forward to the junior officer bunkroom, he thought about Amos Kane and his story.

The word was when Amos had been in flight training, he'd fallen in love with Charlotte, an antiwar zealot. In the midst of a protest outside a US Navy base, a panicked young sailor ran over her with a navy car. The stupid navy cost him his wings, killed the woman he loved, and sent him to a job where he repaired airplanes for pilots to fly. And not one of them was half the stick and throttle jockey Amos Kane had been. Apparently, that's how Amos looked at it.

RT hadn't been sure it was Amos, but thinking about him as he walked, Jon Zachery became sure. Amos sabotaged the planes. Amos killed the two Warhorse pilots shot down, AB and Skunk.

The JOB was well forward, a hundred feet aft of the bow. He ripped open the door to discover Botch, Lieutenant (JG) Butch Felder, sitting on a bottom bunk, not his own. You-caught-me-with-my-hand-in-the-cookie-jar flashed over his face.

Behind Botch, another JG, Tuesday, lay on the sloped bulkhead just below the open porthole. He pushed a bundle through the porthole, followed by a small, shiny metal object. Zachery realized the shiny object was Tuesday's derringer. On combat missions, all the pilots carried a squadron-issued handgun, a .38 revolver. Tuesday augmented his survival arsenal with a personal weapon.

"Hand me the duffel, Botch," Tuesday said.

The bag belonged to Amos.

"Hand me the goddamned duffel!"

Botch handed it up, and through the hole it went. Tuesday bolted the porthole shut, shoved Botch out of his way, and climbed over the bunk to stand on the deck.

"You're not going to rat on us, are you, Stretch?" Botch said.

"Shut up, Botch," Tuesday snapped.

Rake handle thin-and-straight Tuesday and Beefy Botch stood side by side. Stretch looked from one to the other. Botch was worried, Tuesday defiant.

Stretch frowned. "What did you guys—" Then he noted the tattered rug was missing from the deck, and he knew. They had figured out Amos had sabotaged the airplanes and killed two squadron mates. Tuesday had shot him with his derringer. They'd wrapped the body in the rug and tossed it off the fantail. They'd been cleaning up the last bit of evidence of what they'd done when he walked in.

Stretch, Tuesday, and Botch had been good friends. Murder trumped friendship though.

"A note," Tuesday said, "there. On Amos's pillow."

Tuesday brushed by Stretch, took the note, read it, and grinned. "A suicide note. We're in the clear."

Stretch could see Amos as a saboteur but not as a suicide. "You wrote the note," he said and hit Tuesday in the mouth. Tuesday fell back, and Botch caught him.

Tuesday stood up. "I did not write the goddamned note." He wiped blood from the corner of his mouth. "How many times this deployment did you take the law into your own hands, Stretch? Now you're going to condemn me for what I did?"

That hit Stretch harder than any punch Tuesday could throw.

Stretch walked out, stood in the passageway outside the bunkroom, and leaned against the bulkhead. Two questions wanted answers.

"You going to rat on us?

"Now you're going to condemn me for what I did?"

Tuesday had spoken the truth. Stretch had taken the law into his own hands.

His first combat hop, back in November 1970. Stretch was number two in a four-plane. The flight lead was AB. AB was for American Bandstand, LCDR (Lieutenant Commander) Dave Clark's call sign. During the brief, AB said, "Stretch, you're a dumb-shit Newbie. Do what I tell you and not one thing more."

The four-plane worked with a forward air controller (FAC) just below the DMZ (demilitarized zone). AB had rolled in to bomb a cluster of trees the FAC thought might be hiding North Vietnamese trucks. In the bombing run, he came under fire. Jon spotted the antiaircraft gun and rolled in on it. AB radioed for Stretch to abort his bombing run, but Zachery had the North Vietnamese gun in his bombing sight. He wasn't about to abort. Stretch blew the gun up, but AB, the squadron operations officer, was not pleased.

After they landed back aboard the carrier, AB said,

"You were on your first goddamned hop, and you refused to obey your goddamned flight lead—that being me, goddammit. You never duel with a flak site. Everybody knows that. But you!"

AB wanted the CO (commanding officer, also called the skipper) to ground Zachery, take him off flight status for a week, but the skipper refused. He did, however, put Stretch on the schedule to fly with AB for the next seven days. Those flights had been punishment, but despite the punishment, Zachery remained convinced he'd done the right thing—even though he had taken the law into his own hands.

Jon looked up and down the passageway. No one was coming from either direction. He was not close to an answer to his questions.

"How many times did you take the law into your own hands?" Tuesday had asked.

Zachery had an answer to that question: two other times.

During Operation Lam Son 719, the South Vietnamese invaded Laos to cut the Ho Chi Minh trail but got their butts kicked. They retreated and abandoned massive quantities of US-supplied guns, ammo, and vehicles. For weeks, US Air Force and Navy planes flew missions to destroy the US equipment so the North Viets couldn't use it against them. Lam Son 719 was exactly the kind of military operation for which the term FUBAR (fouled up beyond all recognition) had been coined.

On one of those Lam Son missions, Stretch was scheduled to fly with his boss, RT, LCDR Robert T.

Fischer. He was the Warhorse maintenance officer, the fourth senior officer in the squadron and, as such, scheduled to be flight lead for Stretch and two others. In the preflight briefing, RT handed the briefing checklist to Stretch.

Leading a four-plane was unexpected. Zachery had just gotten qualified to lead a two-plane formation.

After the launch and rendezvous, Stretch led his four-plane across South Vietnam and checked in with a FAC with the call sign Oswald.

"Roger, Warhorse," Oswald replied. "Target is an abandoned artillery position atop a hill. No friendlies in the area. Climb to thirty-two thousand feet and hold above the other flights stacked below you. I will get to you in turn."

Another flight leader cut in. "Oswald, this is Raven. I'm bingo fuel and have to return to base. Vacating sixteen thousand feet."

Two other air force flights reported bingo fuel and returning to base.

"Okay," Oswald said. "Who do I have left here? Flights, check in from the bottom."

"This is Hammer. I'm on the bottom, and if you don't work me now, I'm bingo."

"Uh," Oswald. "Hang on, Hammer. I'll mark the target for you with a smoke rocket."

"This is Hammer. Stay out of the way, Oswald. I have the target. I'm rolling in as a four-plane."

"Negative, Hammer. I want individual bomb runs and two runs each plane."

Stretch thought that was the dumbest thing he'd

ever heard. Oswald was trying to control things the same way he would if he only had four planes to deal with. But he probably had thirty-four. And he was running them out of gas, and tons and tons of bombs were being wasted.

Stretch had his formation at twenty thousand feet as he approached the target. He could see planes as dots swirling above the FAC plane like night bugs above a streetlamp.

"This is Warhorse. I have the target in sight. I have Oswald and Hammer in sight. I'm rolling in now with a flight of four. Hammer, I'll be out of your way in seventy-five seconds."

"Negative, Warhorse. Abort, abort."

"Go home, Oswald. Hammer is following navy."

Stretch rolled into a dive and radioed his wingmen to turn on their master arm switches. His dive angle was shallower than normal, only thirty-five degrees instead of forty-five. He adjusted his gunsight. At five thousand feet, he hit the pickle, and his bombs thumped away.

Stretch pulled up and turned right.

"Hammer's in," came over the radio.

As he climbed, he checked his wingmen. They'd hung with him.

"Great hits, navy," Hammer radioed. "Bombs away."

Another flight called rolling in. Oswald remained silent.

Back on the ship, before they entered the intelligence area to debrief the hop, RT told Stretch he had done the right thing.

"Doing the right thing doesn't mean you will get

A TICKET TO HELL: ON OTHER MEN'S SINS

away with it though. You disobeyed a US Air Force FAC. They might want a piece of your hide."

Stretch nodded. He knew. A rivulet of ice water trickled down the underside of his sternum.

But the skipper found out the air force wasn't happy with the FAC either, and, he opined, they did not want to be shown up by a navy lieutenant when there were air force majors holding over that target.

"You were lucky, Stretch," RT said. "This time."

He'd been lucky another time. Most of the Warhorse combat missions were "making toothpicks." Warhorse pilots worked under FAC control, dropping bombs into patches of jungle that *might* be hiding North Vietnamese trucks. Pilots called these "making toothpicks missions," blowing up trees and only rarely causing a satisfying secondary explosion from a munition-laden truck. After such a toothpick mission over Laos, Jon was leading his flight back across South Vietnam to return to the carrier when he was diverted to a FAC. The FAC had discovered Viet Cong running across an open field. After so many useless missions, this one could finally yield a meaningful result.

The FAC talked Jon onto the target. In his strafing run, he lined up his gunsight on a dozen or so black-clad stick figures hightailing it across a field. He squeezed the trigger. Nothing happened. His guns had jammed. He wanted to kill those VC, and he violated a major safety commandment: do not cycle the gun-charging switch. Recycling the switch could ram one explosive bullet on top of another and blow the nose off his airplane and kill him. That didn't matter. Kill lust overwhelmed his common sense and respect for safety rules. He had recycled the switch, and still his guns did not fire.

When he got back to the carrier, his conscience, unable to get in front of what he was about to do, caught up with what he'd done. He went to his CO and turned his wings in. The CO sent RT to speak with him.

"You learned something about yourself," RT had said. "That kill lust surprised you. You didn't know it was in you. Now you know, and you won't ever let that happen again. You're taking your wings back."

He took them.

He thought about Tuesday. Lieutenant (JG) Larry Monday, given his call sign because he was always late. A gregarious, had-an-answer-for-every-question guy who walked into a room and created a top-of-the-social-stratum vacuum—and then he filled it. Normally, such people repelled Stretch, but he liked Tuesday. They were friends. No. He could not condemn him, but they no longer were friends.

Botch was easier to like. He followed Tuesday around like a puppy heeling the pack's alpha. No. He would not rat on them.

RT. He had to get to the ready room and help him.

As he walked through the hangar bay, he thought about Teresa. The kill lust had happened just before Hong Kong. He decided he had to tell her. She had the right to know what kind of man she had married. He intended to tell her the second night there, but he chickened out. As she lay sleeping beside him, he wrestled with the issue and finally convinced himself it was his burden to carry. If he told her, he'd be lifting some of it to her. St. Peter would sort out Jon Zachery's soul at the proper time. He didn't tell her.

Hong Kong passed in a blur of bliss, during which they'd created Little Pootzer. Now Teresa was three months along, and he ached to get back. He loved her always but also loved seeing her pregnant. She just glowed with *God created me to be a mother*. She hated it when her cheeks puffed out, but he always told her he didn't notice. He noticed something else puffed out though. She swatted him on the arm after that one. The recollection flitted past.

Now he had another thing he was not going to tell her. Tuesday and Botch had committed murder. He wasn't going to rat them out, and that smeared his own soul with the sin.

Just before he opened the ready room door, he thought, *The things I can't tell Teresa are piling up.*

Chapter 3

Stretch entered the ready room and stopped. The skipper stood at the front of the room, glowering at him—six footer, broad shoulders, dark hair, brow wrinkled.

"C'mere, goddammit," the skipper growled.

He marched up the aisle between the chairs. "Yes, sir."

"Grab a pad of paper and write up what happened on the mission," the skipper said. "RT and the XO are handling the interviews."

Stretch turned, sat on his second-row chair, pulled out a pad of paper from the drawer beneath the seat, and looked up.

The skipper still glowered at him, and then the glower melted. One corner of his lips curled up. "You got that *Why me?* look, Stretch. You know the answer. You write gooder than the rest of us. Hop to it! Write!"

"Aye, sir."

"Also, I sent Alice (Lieutenant [JG] Mike Allison) to the other ready rooms to have everyone who flew on the photo recce write up what they saw. Include those write-ups in the report. Once you have them all, write up an overall summary, like it would be from the airwing commander."

Stretch frowned.

The skipper grinned.

The man lost two of his pilots KIA, lost two airplanes, and had a saboteur in his squadron, yet he still found Stretch's discomfit something to smile about.

Stretch swung the tray up from the side of his seat, turning it into a desk, then pulled a pen from the shoulder pocket of his flight suit.

At the front of the ready room, Dog Lips sat behind the SDO (squadron duty officer) desk, and the skipper occupied his front-row seat. At the rear, a petty officer manned the counter where pilots signed for their airplanes. Stretch sat in the middle of the room. No one else was there.

Pilot helmets hung from hooks suspended from the overhead. The hooks above AB's and Skunk's chairs were empty. During the current deployment, until today, no one had been shot down. Two F-8s had been lost after their planes struck the ramp at the rear of the flight deck. One was killed. One ejected and was pulled from the water by the rescue helo. Another F-8 pilot died in a flight deck crash during training prior to the cruise. Over an eight-month span, three planes crashed, and two pilots died. Today, in about eight minutes, three planes had been shot down. AB and Skunk were killed. Tiny survived and was rescued.

Tiny was the biggest pilot in the airwing and an F-8

driver, and he was Jon Zachery's best friend on earth. He'd flown escort on the photo plane.

Behind him, the door to the ready room opened and closed. Tuesday hustled up the aisle to the skipper and showed him the suicide note.

Stretch watched the CO study the piece of paper. "Christ Almighty! You know where Kane is?"

Tuesday shook his head. "I just came from the JOB. He wasn't there."

"Give the note to Stretch and tell him how you found it." The skipper handed him the note. "What happened to your lip?"

"Uh," Tuesday said, "I forgot I left the door to my locker open and ran into it."

The SDO's phone rang. He answered. "Skipper, CAG wants to see you."

The airwing commander was called CAG, a holdover term from World War II when the planes assigned to the carrier were called an air group. Now the navy used the term *airwing*, but calling the airwing commander CAW just would not do.

The skipper stood and started down the aisle. Stretch looked up at him.

"Write, I said, goddammit. Hop to it." The skipper stomped away.

Tuesday handed the note to Stretch. Stretch wouldn't look at him.

"You want me to tell you where I found this?" Tuesday asked.

Stretch looked at Dog Lips. The SDO was watching them.

Stretch tore a sheet off his pad and handed it to Tuesday. "Write down how you found it."

Tuesday sat on his chair and pulled up the tray.

Stretch wrote the date and the time on his pad. Then, instead of doing what he'd been told, he started writing the summary. He listed the seven planes committed to the mission, four A-4s, an F-8 photo bird, an F-8 escort—Tiny—and a tanker plane to provide fuel to the F-8s. The F-8s would fly the planned route at high speed and need extra fuel. He described the route that began fifty miles above the DMZ and ran south to just above the DMZ. The four Warhorses provided protection from SAMs. Warhorses One and Two covered the northern part of the route, and Warhorses Three and Four covered the southern half.

USS *Solomons* committed other aircraft as backups for the primary planes and kept another four A-4s loaded with bombs in alert condition in case a rescue mission was required.

"Lieutenant junior grade Kane, report to ready room five" came over the ship's general announcing system. The message was repeated, as it always was.

Tuesday handed Stretch his write-up.

Stretch took the sheet and placed it on the adjacent chair, along with the note. Whatever he'd written was a lie, but it was Tuesday's lie—until he transcribed it into the summary; then it would become Stretch's sin too.

Tuesday left, and Stretch started a fresh sheet of paper and related what happened to Warhorses Three and Four, RT and him.

> Warhorse Three and Four launched first and headed directly for South Vietnam and coasted in just south of the DMZ. The two planes flew west, entered Laos, flew north until they cleared the DMZ, then turned east.

After writing the last, he paused. The US took pains to avoid the DMZ, while the North Viets used it all the time. *If we were fighting the same war the North Viets were fighting, we'd have won it a couple of years ago.* Stretch had had that thought before.

> About two minutes after the flight picked up the eastbound leg of their route, two SAMs were fired at them. Warhorse Four spotted the missiles, and the two-plane evaded them and bombed the launch site.

Stretch included details of the formation they'd flown. RT taught him to fly above and forward of his lead plane. It afforded great coverage of the flight lead to the sides and rear. It enabled the lead to always have his wingman in sight. Nobody else in the squadron used it, but RT liked it. He'd tried to get other squadron pilots to fly combat formation that way, but no one else was comfortable with it. The wingman burned a lot of gas weaving back and forth in front of the lead, but Stretch was sure he would not have seen the SAMs lift off if he'd been in standard formation.

As they attacked the site that fired at them, over the

radio, they heard Skunk report that AB had been hit by a SAM and his plane was on fire and going down. Stretch pulled out of his bombing run and heard the escort F-8 report that Warhorse Two had been hit and was going down also. He observed no chute.

> The SAM site that fired at Warhorses Three and Four had been located in a five-mile-wide valley with ridges on both sides rising to 2,500 feet. Warhorse Three and Four pulled out of their bomb runs, and the valley erupted with AAA fire. Closely packed white and gray puffs of exploding shells appeared to be a carpet strung across the valley from ridge to ridge.
>
> During the SAM and AAA firing, Warhorse Three and Four's warning systems remained silent. The North Viets fired some of the AAA in barrage mode, giving the appearance of carpeting the valley between the ridges. It was highly probable that not all the AAA was radar directed, but some was.
>
> It is apparent the North Vietnamese were ready and waiting for us, and we flew into their trap. After checking my mission chart, I am certain the missile site had been located between three and five miles inside the DMZ.
>
> Warhorse Three led the flight

north and encountered no more SAMs or AAA. Shortly after departing the bombed SAM site, a radio call from the photo bird escort called SAMS lifting off to the west, "Kodak, break right!" Kodak, the photo bird call sign. Some seconds later, Kodak reported his escort had been hit; his plane was on fire but still flying and headed for the coast. Seconds later, Kodak One radioed his escort: "Kodak Two, you're really on fire, man. Eject."

"This is Kodak One. Kodak Two ejected a mile off the coast. Good chute. I got a good chute."

Kodak had also included TACAN range and bearing from USS *Solomons* to the escort pilot in the water, but Stretch couldn't remember those. They'd be included in the write-ups Alice was collecting.

The shootdowns occurred about twenty miles from the start of the photo recce route. When Warhorse Three arrived on scene, the photo bird orbited over the escort pilot. Warhorse Three assumed on-scene command of the rescue effort and sent unarmed Kodak One back to the ship. He also requested the ship launch the alert rescue airplanes as well as two tankers.

> Kodak Two had deployed a dye pack, making it easy to spot him in his raft.
>
> Warhorse Three observed boats pushing away from the coast and heading for the downed pilot. He called the ship to divert any *Solomons* planes returning from bombing missions in Laos and any airborne F-8s to the rescue mission. Then he and Warhorse Four strafed the boats and sank one and disabled another. Two F-8s arrived on scene. Warhorse Three guided them onto the boats. They sank two more. The other two boats went back ashore. Four A-4s returned from bombing In Laos and patrolled at low altitude just off the coast to discourage any of the boats from making another attempt at capturing the pilot.
>
> The helo arrived on scene and hoisted Kodak Two out of the water.

Stretch wrote a separate section about the sabotage. The squadron inspected all their aircraft, and most of them had been sabotaged in the same way. None of the other squadrons' planes had been tampered with.

During interviews with the Warhorse electronics technicians and flight deck personnel, it was disclosed that LTJG Kane was often around them as they worked. Also, he helped them with their tasks, one in particular.

He assisted with testing the electronic warning system on numerous occasions in the past ten days. Sometimes he sat in the cockpit and reported which threat he saw when a tech injected a signal into the system. But he also asked the sailors to show him how to inject those test signals into the system. And when they explained how that was done, he would have seen how cables connected the electronic box to the antennas.

Evidence against Amos Kane as the saboteur piled up. He had the opportunity to commit the crime. His attitude and strong antiwar views provided motivation, and his suicide note constituted a confession. The entire ship had been searched, including the deepest and darkest nooks and crannies, and no sign of LTJG Kane had been found.

Alice and pilots from other squadrons trooped in with a steady stream of statements from others who'd flown on the mission or the rescue of Kodak Two. In addition, Combat Information Center on the ship had recorded the radio frequency used by the photo recce mission. It had been transcribed and gave Stretch a good timeline and verbatim radio transmissions, rather than recollections fuzzied by the heat of battle.

One telling bit was that the warning systems in both Kodak planes had functioned properly. Kodak One and Two both reported two SAMs had been fired at them, which they evaded. Kodak One reported a second two missiles were fired from a separate site. While they evaded those, another missile from the first site nailed Kodak Two.

At 1700, Stretch's stomach growled. He hadn't

eaten lunch. The report was finished, and he took his longhand composition to the admin office and asked a yeoman to type up five copies.

Dinner rejuvenated Stretch. He picked up his report. Alice took copies to the other ready rooms that had contributed to the document. They were asked to review and mark up their copies and return them ASAP. The CO, XO, and RT reviewed a copy and marked it up.

He'd gotten most of the facts correct and most of the inputs concerned formalizing the informal wording in the initial draft. He consolidated all the inputs. At 0100, bleary-eyed Stretch looked over the last rewritten pages. It was done. He returned it to the yeoman for retyping. In the morning, he'd ask RT to review it before turning it over to the CO.

He stood up and yawned and considered going to his bunk in the JOB, but he did not want to sleep in the room with Tuesday and Botch. Stretch was exhausted and practically brain-dead. He understood how horses could sleep standing up. They were as tired as he was. He plopped back down. The seat didn't recline much, but it was a lot better than sleeping standing up.

He thought, *Murderers*. Amos, Tuesday, and himself.

But even murder could not keep the lights inside his head from going out.

Teresa's eyes popped open. Something had awakened her. *The children.*

She heard the knock on the side door, the one leading

to the carport. She pulled on her robe and padded barefooted down the hall and through the kitchen. Amy Allison stood there, also in a robe and wearing fuzzy slippers. Amy and Mike lived next door, their unit attached to the Zacherys' at adjoining carports. Worry, or horror, ravaged her youthful face more than aging ever would.

"Teresa. AB and Skunk were shot down."

Teresa placed her hands over her tummy. "Mike?"

Amy shook her head. "Mike's okay, and so is Jon."

"Oh, Amy." The women embraced.

"Mommy cwying."

Teresa turned. Edgar Jon stuck his thumb back in his mouth. A worried Jennifer stood next to her little brother.

Amy entered the hallway and knelt on the floor in front of the boy. "Sometimes grown-ups cry when they're happy. Grown-ups can be silly sometimes, can't they."

Edgar Jon nodded without removing the thumb.

"Amy, why don't you come in? I'll make coffee, and we can get the children breakfast."

Amy nodded. "I'm going to get dressed first. Things may get to the point where there won't be time later. Then I'll come back. I really don't want to be alone just now."

Amy left, and Teresa closed the door. To her right, the calendar with its circled date accused her. She'd prayed for Jon but not for anyone else.

She hoped Amy would return quickly. Getting dressed was probably a good idea, but having Amy—another adult, another woman—with her held the promise of being able to bear burdens. Teresa thought of Jon. He wouldn't ask for help if his life depended on it. Women, though, sensed pain, grief, and horror in other females. And they didn't even have to say, "Let me help you with that?" No, females just reached out spiritual hands and lifted a part of the other woman's burden. Amy was such a blessing.

"Breftus," Edgar Jon said, fully at home with the imperative.

Chapter 4

The eyes of Jon Zachery's mind opened. Those of his body did not. The open eyes saw SAMs lifting off, saw AAA puffs carpeting the valley from ridge to ridge, saw the cable from the warning system to the antennas with the missing pins, saw Tuesday shoot Amos with his derringer.

The eyes of his body opened. Lights at the rear oozed a thimbleful of illumination into the ready room. Without eyes of any kind, he saw AB's and Skunk's planes going down, trailing flame and smoke, he saw Amos Kane sabotaging their planes, and he saw Tuesday and Botch throwing him into the Tonkin Gulf like a bag of garbage. And he saw his complicity in the latter.

"Damn," he mumbled.

"I don't think I ever heard you cuss, Stretch." RT sat next to him.

Zachery tried not to swear. He hated the notion of having one set of moral behavior at work and another at home. Teresa would never tolerate him swearing in front of her or the children. If he couldn't swear at home, he wouldn't swear at work. And RT, apparently, was the same way. But RT swore yesterday, and Stretch did today.

Zachery leaned forward and glanced at his boss, next to him. In the gloom, he couldn't see RT's eyes, just the dark sockets.

"I'd've said some damns yesterday," Stretch said and flopped against the back of his chair, "only there wasn't time."

"You okay, Stretch?"

He sucked in a lungful and exhaled. "Yes, sir. I'm okay. Considering."

"How come you slept here?"

Stretch didn't look at him. "It was late, really late, when I finished working on the report." He shrugged. "Oh, I kept a yeoman up late too. It should be typed up. Would you read it before I give it to the CO, please?"

"Sure. Right after breakfast. Which doesn't go down for a half hour. Which gives you time to get a shower. You stink, Stretch."

He lifted his arm and sniffed his pit. "A skunk doesn't smell ... Damn."

Skunk was dead. Bits and pieces of him splattered among the bits and pieces of his airplane around a hole in the ground in North Vietnam.

"I stink all right," he said and stepped over RT's legs.

After a shower, after breakfast, and after RT pronounced the report ready to go to the skipper, he said, "Stretch, Skunk was custodian of our classified documents library. Since we are decommissioning after we return to the States, we have to destroy all that material. We'll be in the Philippines tomorrow. I'm giving you Skunk's job. After the skipper signs off on your report, get those documents into burn bags and burn them in the incinerator on base. Petty Officer Twombly in admin can show how to do it."

"Keep busy so I don't have time to think about Skunk and AB. That it?"

"Sometimes, Stretch, I think you might not be as dumb as you look."

RT left. Stretch stayed. There really was no *I don't have time to think about Skunk and AB*. The two of them, and Amos, and Tuesday, and Botch, and himself, he could push them out of the center of his head, but he could not push them far or permanently away. It was sort of like "The Hound of Heaven." He'd read the poem in high school and had gotten an F on the paper he wrote about it. Sister Mathew made him rewrite it three times, until she put a red A on top of it. Then when he was in college, he doubted his faith for a time and came to appreciate the poem more than he ever did that A he got writing about it. "The Hound" was his favorite poem. But what now padded behind him on silent paws was not grace and salvation. Now the hound of hell was after him.

Figuring out the right thing to do was hard enough before. Now he flew an airplane carrying bombs and bullets and capable of killing lots of people.

Get a grip, Zachery.

The skipper walked into the ready room and stopped by Zachery's chair.

"Good," the skipper said. "You're in khakis. Bring

the report, and let's have breakfast in the clean shirt room."

A second breakfast didn't sound like a bad idea.

The skipper led the way. Inside the door to the formal dining room, the skipper stopped and surveyed the space.

Few officers occupied seats at the any of the long tables. It was a nonflying day. In the dirty shirt wardroom, Stretch and RT had been the only diners. He had never been in the clean shirt wardroom before. Tablecloths covered all the tables. It was like the officer's mess on his destroyer when he was an ensign, except on the destroyer, there was one table, with a tablecloth, that accommodated all the officers. The destroyer carried a crew of less than three hundred. The carrier crew numbered more than three thousand.

A dozen breakfasters occupied one table against the rear bulkhead.

There were plenty of empties, but the skipper walked past those. Six at the dozen diners' table had their backs to them, but the others all looked like high school kids. They all needed haircuts and wore wrinkled uniforms. They had to be the Dirty Dozen. All of them were against the war in Vietnam and protested with their unmilitary appearance. Word was the CO of the ship had punished each of them at captain's mast, but he refused to put them off the ship. They were obligated to render service to their country, and by God they would render service, but they'd be watched closely.

The Dirty Dozen. Stretch—and everyone aboard—had heard of them, but he'd never seen them gaggled up like now. When Amos Kane joined them, Tuesday said they were the Dirty Baker's Dozen.

The skipper stopped at the head of the Dirty table. "You sons of bitches, git," he snarled through clenched teeth. Eyes grew big. Mouths dropped open. "Git!"

Chairs scooted back, and they scrambled up and away and out of the space.

The skipper sat. "Nice of them to leave us their table, eh, Stretch?"

"Uh, yes, sir. Nice."

A Filipino steward took their orders and departed.

"You were in admin working on the report last night when CAG visited the ready room. The CO of the ship wanted all us airdales warned. He doesn't want anything to happen to the Dirty Dozen. The CO of the ship personally interviewed each one of the sons of bitches, and he is convinced they had nothing to do with the sabotage of our planes, nor did the sons of bitches know what Amos did."

Stretch wondered if the heavies suspected what had happened to Amos. Maybe the carrier CO was just taking prudent precautions.

"Well, Skipper, nothing happened to them. You just needed this table in the quietest corner of the wardroom to review the report."

"Stretch, you're not as dumb as you look. RT says that about you all the time."

Not all, but many aviators behaved the way the skipper had. Encounter a situation charged with heavy emotional content, deal with it, and then crack a joke. It was as if in dealing with the Dirties, the skipper had dealt with them permanently, as if they no longer existed, and life was a barrel of laughs again. Right now, life was a lot of things but not a barrel of laughs.

The Dirty Dozen. In a big way, they, or the antiwar protest, were the reason Stretch was a navy pilot and had flown that last mission of the deployment. Those filthy, slovenly twelve were anti-America, and so was the protest back home. The night of the dog poop had driven LT Jon Zachery, Stretch, to this place at this time. He'd considered it a call to duty, a noble thing to be here. Now though, he saw his complicity in the murder of Amos as undermining to the root principles of the nation as was the protest. And he didn't know how to deal with the notion.

"You going to eat your omelet?" the skipper said.

Stretch picked up his fork and started eating.

The skipper snorted. "After you finish your breakfast, come up to CAG's office. He may have a few questions about something in your report."

The skipper stood, and Stretch jumped to his feet.

"Oh, for Christ's sake, sit down and eat."

Stretch stood at attention.

"Judas fribble-frapping priest!" the skipper said. Then he walked away behind Stretch.

Stretch didn't turn but stayed at attention as he counted, *One potato, two potato,* up to twenty of them. Then he sat and ate. Denver omelet, and it was a good one. Bacon, always good. Well, not raw. Wheat toast, good. Coffee—weak-weenie, surface navy stuff. CAG would have some good, nasty, black, melt-your-teeth-enamel stuff.

Stretch stuck the last bit of toast in his mouth, stood, and wiped his lips with his cloth napkin. On his destroyer, the XO would have counseled him, "Dab your lips before you stand. You want people to think you're an aviator?"

The right and wrong of things was a moving marker. In the dirty shirt room, a guy was as likely to use his sleeve as the paper napkin. Time was important. Get your food. Eat it. Get back to work; and decorum, if you have time, well, go for it. Otherwise, *get back to work* was the guiding light.

His thoughts leapt from napkin etiquette to: *thou shalt not kill.*

In the literal sense, it did not apply in combat. In combat, it was your duty to kill the enemy of your country.

He sure wanted to kill me.

For Jon Zachery, the enemy in Vietnam had a face. It was 1966. Jon's destroyer, in company with another, steamed north into the northern part of the Tonkin Gulf. Two North Vietnamese PT boats attacked the US Navy vessels. Both PTs were sunk by navy aircraft. Jon's ship put a boat in the water to rescue a North Vietnamese survivor. Jon was the boat officer. As one of his boat crew reached for the North Vietnamese, the enemy lunged at the would-be rescuer and stabbed him in the shoulder. Jon could still picture the hate on that young North Vietnamese man's face. He could also see himself pull his forty-five, rachet a round into the chamber, flick off the safety, and fire four rounds.

One hit his target's shoulder, one the side of his head, and last two went into the water after the man sank.

The first two rounds he fired, the ones that struck the man, sat okay, if not comfortably in his conscience. The last rounds fired into the water, it was as if he'd absorbed his enemy's hate and fired at the water, and fired again, overcome by kill lust. The face of the first man he'd killed came out of the box in the attic of his mind for frequent viewing. Zachery always wound up wondering what his own face looked like when he fired that forty-five. Was his the face the Viet Cong and North Vietnamese pictured when they thought about the US enemy? For him, there was no question. That PT boat sailor's face was the face of the enemy.

Stretch felt eyes on him. A table occupied by a half dozen officers, one of them a commander, stared at him from one table away.

LT Zachery popped to attention. "Good morning, Commander, gentlemen." As he left, he felt their eyes follow him all the way across the wardroom.

In the airwing commander's office, CAG wanted to know if Zachery was sure the North Viets fired SAMs from within the DMZ. Jon hustled down to the ready room, retrieved the map he'd marked up for the mission, and hustled back. His map was marked with TACAN ranges and bearings at points along the route RT and he had flown.

"The route we flew across South Vietnam was exactly as we planned," Zachery said. "I ticked off a couple of the TACAN points as we headed west. We entered Laos and turned north, again, exactly on the

TACAN coordinate we planned. We turned east exactly on the planned turn point. From then on, I was too busy to look at my TACAN, but I can tell you the place the SAMs came from is right here on the chart. See these elevation features rising to seven and eight hundred meters on the sides of this valley? That valley is where the SAMs came from. At least three miles south of the northern border of the DMZ. Maybe five."

"Okay, Lieutenant," CAG said, and he called for a yeoman, then told him to type up the summary section of the report as a message.

Outside CAG's office, the CO said, "RT said the same thing you did."

"So, this was a test, to see if our stories were the same?"

"Stretch, in the heat of battle, details get fuddled up. You and RT knew what you were doing. You had a good plan. You executed it. But CAG had to ask."

"We had to prove we were innocent."

"Stretch, you're not as dumb as you look. RT says that about you all the time."

The skipper walked away. Stretch leaned against the bulkhead. The report was finished. As he'd worked on it, taking inputs from all the sources and consolidating them, he'd felt as if he was on a speeding train, and the world outside the window moved past as blurred features. Now, suddenly, he'd found himself standing on the side of the tracks, and the train sped by him. It disoriented him.

"A suicide note," Tuesday had said. "We're in the clear."

The report he'd written set that in concrete.

Lunchtime. Then he'd have to figure out the scope of the job of destroying the classified library.

Teresa answered her phone in the kitchen. Helen Fischer, RT's wife, said, "The XO's wife asked us to fix dinner for Sybil and Lauren."

AB's and Skunk's wives.

"I said you, Amy, and I would take care of Sybil," Helen said. "I hope that's okay."

"Of course," Teresa said. "I can make a chicken enchilada casserole—oh, Sybil. She'll turn her nose up at a casserole."

"I'll take care of the main course. Can you and Amy make dessert, a salad, and green beans?"

"Sure. I'll talk to Amy."

"Thanks, and how about if I come to your place at four thirty, and we can take it to Sybil then."

Teresa hung up. She was sure the other wives all wanted to help Lauren. Nobody wanted to see Sybil's grieving widow act. Bless Helen.

Jon admired Helen's husband, RT, and Teresa admired how Helen always stepped up to do the toughest or most unpleasant jobs. RT was leaving the service after the ship returned, and Jon had written that he'd miss working for and flying with him. She'd miss Helen.

Jon had also written that RT would have fifteen years of service when he left, five short of earning a pension. But RT wanted to have a *normal* life and to be present for his children as they grew up.

It would be nice, Jon Zachery, if you felt that way about the children and me.

After the ship and the squadrons decommissioned, Jon expected to be sent to a meaningless job, but she knew he'd be scheming to get back to the war as long as it continued. And it was hard to see an end to it.

Teresa walked to Amy's back door, knocked, and told her neighbor what she had been volunteered for.

"Sybil." Amy shook her head. "A nasty job, but someone has to do it. That it?"

Chapter 5

"Mr. Z.," Petty Officer Twombly said, "before we burn the classified library, we have to inventory it. Also, the secret and top secret stuff, we have to page count that."

Twombly was a YN2 (second-class yeoman). He held a top secret clearance, and he was the assistant CMC (classified material custodian). He showed Lieutenant Jon Zachery a document naming LT Zachery as the Warhorse's CMC. It was signed by the executive officer.

Under his blond crewcut, the YN2's blue eyes sparkled. "Congratulations, sir."

The new CMC humphed. "The inventory. You have a list of all the documents, I presume."

Twombly handed over a stack of pages stapled together. Altogether, twenty-five pages with single-spaced entries, but at least entries were typed on only one side.

"You have all this stuff?"

"I have some of it, sir. But a lot of it is checked out. The officers have some documents, and electronics technicians have others describing the function and maintenance procedures for some of the equipment, like the jammers in your airplane."

"And the threat warning system."

"That one too, sir. The ordnance men have other documents. The CO, XO, the ops O—Twombly paused over AB's title—and several of the other officers have things checked out."

"Let's start with the things you have in your custody," Zachery said.

Some of the items were books two inches thick. Some messages and memos consisted of a single page.

Confidential—the lowest level of classification—material was kept in locked file cabinets in admin. About one third of the confidential items were in the cabinets. The library checkout system consisted of a box of three-by-five cards listing the document, who signed it out, and the date. They checked the inventory and found only half the missing items had a sign-out card.

"What kind of a system did you run here, Twombly?"

"Uh, Mr. Zachery—"

"You don't want to say anything bad about Skunk, Lieutenant Kohler. Because he was shot down, right?"

"Mr. Kohler got a little, uh, careless the last couple of months. He said, 'What the hell does it matter? We're decommissioning as soon as we get back.' I told him we were going to have to document the destruction of all this stuff, and I told him I wasn't going to sign any paper lying about whether we found a particular document or not."

"What did he say to that?"

"Nothing. He just walked away."

Poof! Stretch's sympathy for Skunk started disappearing, but he grabbed a handful before it got totally away.

"Okay, YN2, here's what we'll do. You take the checkout cards and see if you can round up all that material. I'm going to start with LCDR Clark and LT Kohler. You have the combination to the safes in their rooms, right?"

Twombly went to one of the locked file cabinets and rifled through a handful of sealed envelopes and handed over two of them.

"Then," Zachery said, "I'm going to talk to all the officers and see if they have any of the unsigned-out material. Let's meet back here at seventeen hundred and take stock of where we are."

Six confidential items were missing. Zachery sent Twombly to chow. He found the skipper in the dirty shirt room and told him about the missing items.

"Both AB and Skunk had a lot of confidential and secret stuff in their safes," Stretch said. "I'll bet they stuck some classified items in a drawer with their skivvies."

"I wasn't going to have their personal effects inventoried until we left the Philippines. Nothing much to do while crossing the pond. I'll have the XO get some guys to do AB's and Skunk's gear."

"Amos," Stretch said.

"Amos goddamned Kane. I haven't thought about

him since CAG sent the report out. I guess my mind was anxious to get him out of there. Okay, I'll have the XO tell Tuesday and Botch to do it."

"I'll do it, sir. Alice will help."

The skipper stared at Stretch for a moment. "Okay, then. I'm going to cancel the movie for tonight. I'll tell the ready room it was your idea."

"Well then I'm going to need to check out my .38 again."

The squadron used .38-caliber revolvers as survival weapons. All had been collected and turned in to the ship's armory at the end of combat ops.

The skipper stared again. The enlisted men talked about that stare when the CO had them at mast for some violation of military rules. They dreaded the stare more than the inevitable punishment to follow.

"Stretch, as I recall, you told RT that sometimes it seems like your mind has a mind of its own. You told him that after you recharged your guns, right?"

He nodded.

"I thought that was reasonably close to halfway insightful. It's worth remembering, and I thank you for it. So, here's a variation on that theory. Sometimes our mouth has a mind of its own." The skipper looked at the food on Stretch's tray.

Stretch snatched up his drumstick and took a bite.

The skipper left.

Stretch's mouth decided to stop chewing.

He'd slipped into a contest of snappy repartee with the CO, and these were always jousts where the winner got in the last word. His comment about checking out his .38 was the last word, so he'd won. But in this contest, in winning, he lost more than the loser did. There'd been enough Warhorses killed even if self-defense applied.

Rather than cancel the movie, the skipper delayed it by two hours.

The missing confidential documents were all recovered. RT and Skunk had five between them. Amos had one, a maintenance manual used by the ordnance techs. Part of this document discussed the bomb-arming system in the A-4. In addition to the book, Stretch and Alice found a wire cutter, sections of arming wire, and a bomb-arming solenoid. Apparently, Amos had experimented with snipping the arming wire only partially through. Apparently, he'd figured out how to dud the bombs without leaving evidence of tampering visible to the sailors and pilots who checked the weapons prior to launching.

Amos Kane had been responsible for the dud bombs. Zachery saw no reason to insert an *apparently* in that thought.

It was midnight when Stretch returned to the dim lit ready room. The movie was over, and the place was empty except for an enlisted sailor manning the SDO desk. A desk lamp cast light on him.

"Evening, Petty Officer Astor," LT Zachery said.

Astor was a squared away sailor. He ironed his dungarees, even at sea. Everyone called him JJ, for John Jacob, though his name was Herman. He returned the greeting.

Stretch sat, pulled a pad of paper out of the drawer, and converted his chair to a desk.

"Yesterday was a hell of a way to end this deployment, wasn't it, Mr. Z.?"

He was about to write *Dearest Teresa* on the paper. Instead, he flicked off the red-lens flashlight.

Astor was a good kid. He worked hard on the flight deck. *It must be hard to put in all that work and not see what it accomplished, or if it did any good at all.* For that matter, what good did a logbook full of *making toothpicks missions* do?

"Well, PO, I can't argue with what you said."

"It's like we got our butts kicked, sir. I mean it's like we're going home like a whipped puppy, whimpering and our tail between our legs."

"Is that what you think?"

"I don't know what to think, sir. I heard some of the guys talking though."

"You know what Lieutenant Kane did, right?"

"Everybody knows, sir. We all thought he was a good guy."

"The North Viets set a trap for us. It wouldn't have worked so well if Kane hadn't sabotaged the airplanes. As it was, we got in a lick or two of our own. LCDR Fischer destroyed the SAM site that fired at us. So, no, I'm not whimpering, and my tail is not tucked between my legs. The thing that really crimps my skivvies in a knot is that we are decommissioning after we get back. I sure wish we'd come back for another whack at them."

"Thank you, sir. Sorry I bothered you."

"You didn't bother me, JJ."

"You know they call me that, sir?"

"Everybody knows, JJ, and we all *know* you're a good guy."

JJ's face lit up, and with his smile still blooming, he cracked open his paperback to the dogeared page.

LT Jon Zachery, master of snappy repartee, had given a sailor something to smile about. It seemed like the first worthwhile thing he'd done in a long time.

He wrote *Dearest Teresa* and waited for the words to come, and waited, and waited.

Whispering woke him. The clock on the forward bulkhead said 0350. Another petty officer was taking over the watch from Astor. Stretch looked at his pad of paper. Only the salutation and the date had been written there.

His brain buzzed. His eyes burned. In his mouth, it tasted like he'd bitten the head off a match.

A shower and a clean uniform restored him. He brushed his teeth, tongue, and the roof of his mouth and savored the taste of Colgate.

In squadron admin, he opened the secret drawer and ticked off the contents against Twombly's inventory sheet. All present and accounted for, which was good. Missing a piece of confidential was one thing, but it was altogether another to lose a secret document. He nodded to Skunk for not totally doping off on the job.

It occurred to him that people in normal jobs, when they died, left a putrefying body for someone to deal with. In his own job, when people died, they left their bodies at the bottom of an ocean or in a hole in the ground in North Vietnam, but they left other vile and stinky things for people to deal with, as Amos did.

He thought about Teresa. The reason he couldn't find words to write to her was because he hadn't figured out how he was going to handle the murder of Amos. It was a matter for the confessional. But confessing it, and receiving absolution, removing the sin from his soul, that was like Pontius Pilate washing his hands of the matter of dealing with Jesus. The Jewish people who came to him wanted Jesus put to death. PP knew it was wrong to do so, but he washed his hands and walked away.

Zachery's complicity in the murder of Amos was not a thing he could shuck off like that. The problem with Teresa was she was so insightful. She would smell the murder on his soul over the telephone when he called her. And he had to call her after the ship pulled in at noon.

He recalled the kill lust incident. She hadn't sensed that in Hong Kong. Then he knew what he'd do.

The phone rang, and jerked Teresa from dreamless sleep to heart-pounding fear. It took a moment for it to register that the children were all right, that she was all right. Her hand touched her chest above her heart, then it grasped the phone, and she answered.

Jon. The clock showed 3:00 a. m. But it didn't matter. After finding out about AB and Skunk being shot down on the last day of combat ops, well, being awakened and scared half to death, it just did not matter. Jon had not been shot down. Jon was okay. She lay on her back, held the phone to her ear with one hand, and placed the other on her tummy bump.

"Teresa. Teresa. Are you all right?"

She huffed out a breath. "Yes. Now that I hear your voice. We heard about AB and Skunk, and I knew you were not hurt, but ... it wasn't real until I heard your voice."

Teresa wiped tears from her cheeks.

"It's so good to hear your voice, Teresa. I mean, it grounds me."

"Losing two friends, it must be awful."

Jon didn't answer for a moment. "Yes. We've lost pilots before. The F-8 guys who crashed, landing aboard. But it was always in the middle of operations. After a crash on a carrier, it's like 'Sweep that debris off the flight deck. Pilots, man your planes. Launch the next go.' There isn't time to even say, 'Poor Ham Fist,' or whoever. This was so different. On the last flight. It just seems so ... unfair."

Always before, they limited their overseas calls to twenty minutes. A twenty-minute call was expensive, but the real-time connection was worth so much more than the money. However, when the clock said 3:15, Jon began to say goodbye. Teresa was disappointed, but then, talking about Skunk and AB had to be hard for him.

She hung up and lay there staring up at the ceiling, and it hit her. Jon would have another tragedy to deal with after he returned. His squadron would decommission.

He was right back in the position he'd been in a year ago, in a unit slated to disband, and he had little chance of landing a good job—at least one he would consider good. To him, another flying job and another combat tour to Vietnam would matter.

Teresa looked at it in another way. Jon had his opportunity to fly combat missions. He'd lived through it, and she was overjoyed that he would go to some dead-end job. He'd serve out his obligation, and then they'd leave the service. Their new baby would grow up with his, or her, father at home every day and every night. Not like Jennifer and Edgar Jon had experienced. Their father had been letters and audio tapes.

Teresa asked the Lord to help Jon accept what would happen to him and to see the blessing in it, as she did.

Then she got up, found her slippers, and went to the children's room. They were both sleeping. Carefree and filled with peace and innocence. "Daddy's coming home," she whispered.

Back in bed, Teresa sighed. It would still be a month before Jon got home. Crossing the big puddle of water took time.

It will be so good to have you home, Jon Zachery.

Better still was the rosy future she saw for her family. Jon would serve out his time, be done with combat, done with flying off aircraft carriers, and done with flying period. It was the first time since 1966, after the night of the dog poop, when Jon told her he was applying for flight training, that she could think about a future that did not include consideration of widowhood.

Chapter 6

Zachery left the phone exchange and returned to the ship.

Solomons was tied to the carrier pier adjacent to Naval Air Station Cubi Point. A steady stream of sailors and officers filed off the fore and after gangways. The officer and enlisted clubs would do a booming business that day and the next two with the carrier in port. After crossing the quarterdeck, Stretch went to the ship's chapel and sat in the back row.

It was a compartment with a folding table for an altar and folding chairs for pews and room to seat twenty-four. The chapel easily accommodated those who attended daily Mass or other denominational services. Hundreds attended on Sunday morning. On Sundays, ministers used the forecastle, a large, open area under the flight deck all the way forward. Worshipers had to step over and around anchor chains, but it was out of the way of the never-ending work on airplanes and the equipment used to launch and land them. On a carrier, when the planes and equipment were not being flown and operated, the planes and equipment were being worked on.

Sounds of that work penetrated the space but didn't seem to disturb the solemn silence abiding there. The sounds came from far away, from the moon.

The chapel looked like other spaces on the ship. Visible I beams supported the overhead and framed the bulkheads. The bulkheads and overhead were gray. Green tile covered the deck. Not an ounce of effort had been spent to ameliorate the bleak appearance. Jon had not come to the boxy space looking for stained glass windows, gothic arches vaulting the ceiling, and statuary. He had come for the solemn silence.

Father God, who art in heaven.

That was how Teresa prayed sometimes, and he latched onto it for the comfort it contained in the attachment to her.

Father God, I am right back where I was ten years ago as a freshman at Purdue. Have mercy on me.

During the second semester of college, electrical engineering student Jon Zachery doubted his faith. After growing up with devoutly Catholic parents, after Catholic grade school and high school, now he asked, *Where is the scientific evidence there is a god, any god?* And of course, wasn't religion the opiate of the masses? Jon Zachery, with his superior college-freshman intellect, was so far above the masses in learning and powers of discernment it was stupendous, and it was also the most miserable year of his life.

Teresa picked up on it in their phone calls. "What's the matter, Jon? Lately, you seem sad. You've always been happy, and you lifted me up. What's wrong?"

He couldn't tell her. Teresa was deeply religious.

If she knew he questioned the existence of God, she'd leave him.

"I miss you," he'd told her. It was the honest-to-God truth, whether there was one or not. "I wish we could get married now, but I can't support you, and certainly not a family, on third-class petty officer pay. I figure I have a shot at making second class next year. Then we can get married. I just wish we didn't have to wait another year."

"I wish we could get married now too, Jon. But your navy scholarship is a gift from God. The navy is paying you your regular salary to go to college. And it pays the tuition and for your books! Do you not see the hand of God working here?"

He said he hadn't seen it until she pointed it out for him. Then he dug the pit of his lie deeper by thanking her for helping him see the truth of what she'd said.

That semester, he was taking a required psychology class. Hard science engineering students apparently needed some exposure to softer social science. One lecture dealt with the idea of deferred gratification, which the professor explained as the salvation of the middle class. Put off having a family until completing college and landing a good job. Deferred gratification sounded like an opiate for the middle masses.

He didn't tell Teresa the last part. He did tell her about the lecture and that "Deferred gratification stinks to high heaven."

She'd laughed and said she was happy to hear him *happy* again.

Except he wasn't.

There was no god, but there was a priest. Jon confessed his misery and the source of it. Jon waited for the priest to say, "My son." If he did, Jon would leave the confessional.

"Ah, my fellow sinner," the priest said. "You are miserable because God was a big part of your life. You put Him out of it, and now you suffer for the lack of something vital, like oxygen for your soul."

It turned out the priest, Father Dominic, had questioned his faith too, when he'd been in seminary no less. "What saved me," Father Dom said, "was discovering the power of silence. Especially when you mistakenly think you've lost God. The most important thing is to find some profound quiet and to stay in the stillness patiently until some of the quiet outside oozes its way into your soul. Then you will find you didn't lose God. You will find He was there all along, waiting patiently for you to rid yourself of the clamor of living in our world."

In college, Jon found the kind of silence he needed in the chapel when no one else was there. Father Dom had been right. Jon discovered that God had not left him, that He had been there all along. It was just that Jon's head had gotten filled with a clamor of his own manufacture. Faith and two thousand years of Catholic Church history were nothing compared to the superior intellect of one Jon Zachery, and Jon Zachery wanted physical proof God existed.

In college, in the chapel, he'd knelt for a while, but he kept checking his watch. Sometimes not even a full minute elapsed between checks. And he thought, *Okay,*

A TICKET TO HELL: ON OTHER MEN'S SINS

God. Here I am in the quiet. Where the heck are You? Do You think I have all day? Then he remembered *stay in the quiet patiently. Patiently.* Perhaps if the priest had underlined that word, it would have sunken into his superior intellect more readily.

Jon had removed his watch from his wrist and stuck it in his pocket, and he sat and allowed time to just slip away. Lo, it happened just as Father Dom had described it happening to him in seminary. God was there all along, waiting so patiently it was as if He had eternity at His disposal or something. God appeared like something you see, or think you see, out of the corner of your eye, and you wonder if you've seen a ghost. He appeared like that, and Jon wondered, *Is it You?*

And God flooded in like a burst reservoir dam. Emotion welled up out of his belly. He almost cried, but Jon Zachery was no sissy. His eyes did secrete a little more moisture than usual. And there was his nose secreting its own *ucky* stuff. Thank God for handkerchiefs.

I'm glad You're back.

I never left.

That was how it happened in college. Now on his back-row chair, in another chapel, Jon looked up at the crucifix hanging on the forward bulkhead and waited patiently. What came out of the silence was:

It was all a waste.

Everything since the night of the dog poop was a waste. The protestors were trying to tear down the structural foundation of our country. He disagreed and had to demonstrate, to stand up for what he believed in.

But when we—Tuesday, Botch, and me—killed Amos, we did exactly the same thing the protestors were trying to do. Undermined the structural foundation of the country.

The logic and the moral path he followed led him to: *Tuesday was right. I took the law into my own hands.*

And that, he concluded, he undermined the oath he took to defend the Constitution.

Father God, I know You're in here. I just can't hear You with this other stuff yammering at me like a rivet gun.

And the other thing. Calling Teresa. You know how much it helped me to hear her voice, and You know what I did, called to catch her asleep in the middle of the night so she wouldn't smell the sins on my soul. I don't like me very much right now.

Thanks for being here. Maybe tomorrow I'll be able to listen better.

"Thanks," he whispered and then left, closing the door softly.

He stood there for a moment in the passageway, not sure where to go or what to do when he got there.

"*Solomons,* departing," came over the ship's announcing system. It meant the CO of the carrier was going ashore. Everyone not on duty would be on liberty.

The chapel was on the 01 level, one deck above the hangar deck. Skunk's room was third-deck amidships. There, Zachery knocked and received no answer. Skunk and Dog Lips roomed together.

Next, he went to the ready room and asked the petty officer on duty if the XO was aboard.

"No, Mr. Z.," the kid said. "Everybody's on liberty but you and me."

He used the SDO's phone and called the ship's supply department and asked the duty steward to bring a key to Skunk's stateroom.

Back on the third deck, he knocked again, and again received no answer. With the key the steward brought him, he let himself in.

Skunk's gear had been packed in a metal cruise box, which sat against the bulkhead at the foot of the bunks. Skunk's safe stood open. His drawers were empty.

It took Zachery four trips between the JOB and Skunk's room to transfer all his gear. Once that was done, he set a new combination to the safe door. Then he turned out the overhead light, got out a pad of paper, turned on the lamp above the fold-down lid desk, and wrote *Dearest Teresa*.

He sat in the bubble of light from the desk lamp in the dim room, in chapel-like silence, outside noises coming from the moon, and ink flowed from pen to the page, to Teresa across ten time zones.

He wrote about the Cupid Day. Near the end of their junior year of high school, after attending classes with her for three years, and neither of them more than a casual acquaintance to the other, he'd seen her walking home after school let out one day. He was in the back seat of Harriet Winston's convertible. As they drove past, Jon turned to look at Teresa, and it happened. Cupid shot him in the heart.

Recounting that in intimate and elaborate detail filled a page on one side. Then he wrote about getting his class ring and never putting it on his own finger but giving it directly to her.

He wrote about going steady with her through his first two years in the navy and dreading mail call every day because so many of his acquaintances received Dear John letters, and if he ever opened a letter beginning with Dear Jon instead of Dearest Jon, it would rip his beating heart from his chest before he could read the words.

He wrote about falling ever more deeply in love with her during their senior year in high school. How it was like falling and riding a rocket into the heavens and both feelings burning inside him at the same time and neither of them contradicting the other.

He wrote about falling in love all over again, so many times, and each time his love for her took on a new aspect, as if she were a diamond, and at each new falling, the master diamond cutter in heaven showed him the latest facet He'd carved on His masterpiece.

His pen paused. He stared at nothing and, at the same time, stared at everything. Sitting in his bubble of light in a universe of darkness, she was in such a bubble too, far, far away yet at the same time so very near. Like rocketing up and falling at the same time, violating no law of physics in the realm where love lived. It was as if his heart was a kettle and he'd just dumped the entire contents of it on the pages in front of him.

Except that last wasn't true. There were things he couldn't tell her and maybe never would.

That engendered a need to confess something.

His eager pen bled about questioning his faith in

college and how miserable it made him until the priest, Father Dominick, and Teresa helped him find that he hadn't pushed God away. Nobody could do that. God was always there, but sometimes a person had to find profound silence to discover He'd been there all along.

He confessed he knew she expected him to give her an engagement ring at Christmas his first year of college, but Jon's father had terse philosophies doled out on appropriate occasions. *Neither a borrower nor a borrower be.*

He confessed that when Petty Officer Zachery reenlisted for six years to attend college, he'd received a reenlistment bonus and spent it on a car. It was a six-hour drive from Purdue to St. Charles, Missouri, where Teresa lived. He planned to come home to see her once a month. It took every penny he had to do so. The navy paid for a lot, but he had to cover room and board. And trips home.

Once, during a weekly phone call, during his second year of college, she'd told him she thought they should both date other people to make sure they really were committed to each other. He'd considered the misery he'd experienced after doubting his faith as the supreme misery. *Date other people* recalibrated his misery meter. Their weekly phone calls continued, and each new date she told him about ratcheted his pain above the full-scale peg. He was sure it was preliminary to, "Jon, I've fallen in love with Herbie Smedlap. We're getting married."

Worrying about getting that call drove him to contemplating suicide. It also drove him to a jewelry store in Lafayette, Indiana, where he worked out a payment plan—sorry, Pop—and ordered a ring. An off-the-shelf one would never do. It took two weeks to fabricate the design, two misery-filled weeks anticipating Herbie Smedlap giving her his ring first.

But Jon showed her the ring at second-Purdue Christmas and asked the question.

"Of course, I'll marry you, Jon Zachery. But we've been going steady for four years. I do not want to be engaged that long."

He wrote: *That, as I recall, was the first of many times I said, "Yes, dear."*

The door to the stateroom opened and slammed against the bulkhead.

"*Tha'sh* Skunk's desk! Get outa here, shithead."

Dog Lips reached for the light switch, missed it, and staggered into the room. With his second stab, the lights came on.

Hanging onto the doorjamb, Dog Lips spun around. "This is my room, you little shit. I don't want you in it. Get out."

"You're drunk a—"

Dog Lips was six two with broad shoulders. His thin, dark lips quivered.

"Don't you say his name. Now get out."

"I'm staying."

"No, you're not."

Dog Lips lurched forward and reached out a big paw to grab Zachery. Jon backhanded the paw away, leapt to his feet, and rammed his left into the big man's belly. Dog Lips staggered backward, grabbed the doorjamb, and power puked into the room.

The smell could have come from the belly of a

vulture. Zachery's stomach considered contributing to the mess on the deck.

Dog Lips glowered. "You made me puke. You clean it up. Then get the hell out of my room. I'm going to get the XO. You best be gone when we get back." He started to turn away, stopped, and pointed. "That best be cleaned up."

He grabbed for the doorknob, swore, and slammed the door.

Zachery flopped back on the chair and coasted down from the rage that erupted out of his belly when Dog Lips busted into the room and jerked him from placid paradise to ugly present and puke.

He looked at his letter to Teresa and sensed how far away she was now. A couple of heart beats ago, she was as near as his left hand—the hand that made Dog Lips puke.

One of Father Dominick's sayings popped to mind. "In Corinthians, Paul writes about faith, hope, and love, and the greatest being love. Paul knew some stuff, not only how to use commas, but he didn't know everything. There are times when there is something greater and more important than love, and it is humility. Do you see that, Mr. Zachery?"

He hadn't had an answer for Father Dominick, but he had one for himself now and looked at the elongated puddle of puke.

Judas priest! He must have eaten four orders of escargots.

Zachery wrinkled his nose, got up, took two of Dog Lips's civilian shirts from his locker, and did the initial mop-up.

Chapter 7

Jon woke in Skunk's bed, to the smell of bleach lingering from the cleanup of Dog Lips's mess.

The XO had not showed up last night. He figured the odds of the XO showing up at that late hour on the first night of liberty were 10 percent. The XO would have told inebriated Dog Lips to come back in the morning. Jon expected his hew roommate to come back though and maybe even puke again, but he hadn't. Maybe, he thought, he slept in the ready room, and maybe he puked there.

After completing his morning business, he stopped by the ready room, and the SDO (squadron duty officer) told him that last night, Dog Lips had fallen down a ladder and received a concussion and a broken leg. He'd been taken to the hospital on base.

The enlisted SDO said, "Fly combat missions for six months without getting a scratch, and after one night of liberty, he gets busted up bad. Bummer. But shit happens, right, sir?"

Stretch nodded, left, ate breakfast in the dirty shirt

dining room, and went to admin to meet Petty Officer Twombly to burn the classified library.

PO1 (First Class Petty Officer) Wazinski ran admin. He greeted LT Zachery. "Sorry, Mr. Z., but Twombly's majorly shitfaced this morning. You want me to put him on report?"

"He's a good guy, and he works hard, right?" Zachery said.

"Yes, sir. At sea, he does the work of five men. We hit port, and he drinks the booze of five."

"One of your other guys willing to stand in for him and help me burn the classified material?" Zachery said.

A PO3 at one of the desks said, "I'll do it, Ski."

Zachery nodded to the sailor. "One other thing," he said to the PO1. "Hopefully we're back in time for noon chow. I need Twombly to help me with the top secret stuff. If he's ready to do that, he'll get no grief from me."

"He'll be here," Wazinski said. "He may not look like much, but he'll be here."

Zachery and the PO3 were back in admin at eleven. Twombly was there, seated, his head propped in his hands, his elbows on the desk. He looked up.

"Jesus, Twombs," the PO3 said, "you're bleeding to death through your eyes."

Twombly groaned.

"Petty Officer Twombly, did you have anything to eat?"

"No, sir. I'da puked."

"PO1, can you get him to chow, please? Make him eat some fried potatoes. Then pour black coffee in him. Take a bucket with you, just in case, and have him back here at 1245."

Wazinski nodded, and Jon departed for the dirty shirt dining room, which was packed, unlike at breakfast.

Jon found a place next to the XO.

Normally, a squadron XO would be a commander and slated to take over command in a year and a half. Since the Warhorses were scheduled to decommission, a lieutenant commander was deemed adequate. And that's how the guys in the JOB judged the XO. Adequate. Judge told himself, *Judge not*. But he did.

"Did Dog Lips talk to you last night?" Jon said.

"Sure. At the O Club. The last thing I told him was to go back to the ship, that he'd had enough to drink for one night. Guess I was right. Guess I should've sent someone with him."

"Heck of a thing, but did you talk to him after you got back to the ship?"

The XO shook his head. "He'd already fallen down the ladder and was in the base hospital before I got back. Why?"

"While everyone was in the club, I moved into Skunk's half of the stateroom."

"And Dog Lips didn't want you there, right?"

"He did not. Apparently, he was coming to see you when he fell."

"You moved in without asking Dog Lips or me? You just moved into Skunk's bunk? You remember Dog Lips was supposed to go on the last hop, but his plane went down, so they launched the spare, Skunk?"

"I remember, sir."

"I'm surprised Dog Lips didn't just throw your ass out."

"He's a big guy, sir, and I don't know what kind of a fighter he is sober. Last night, all he could do was puke at me." Zachery dropped his eyes to his tray, then back up. "I have to burn classified material this afternoon, sir. When I get back, I'll move out of Skunk's place."

"Back to the JOB?"

"No, sir, I'm not going back there."

"Oh, for Christ's sake! Stay in the goddamned room. I'm getting out of the navy in three months, if one of you goddamned JOs don't give me a goddamned fatal heart attack first."

Zachery dumped his full tray and returned to admin.

The three admin POs showed up on time. Twombly's dungaree uniform had been ironed. He'd shaved. He looked squared away, as long as you didn't look at his eyes or his gray pallor.

"He puke?" Zachery said.

"He said *urk* a couple of times," Wazinski said, "but he held it in."

"You ready to go to work, Petty Officer Twombly?"

They filled two bags with top secrets and took them to the base incinerator. After disposing of them and signing the destruction sheet, Zachery drove them back toward the ship.

"Do you jog, Petty Officer Twombly?"

"No, sir."

"This afternoon, you will jog with me, and that's an order."

"I don't have any running gear, sir."

Zachery turned into the base exchange parking lot and grinned. "We can fix that."

"Shit," Twombly said.

After Zachery paid for the gear, and as they drove the rest of the way, Zachery said, "Can I give you some advice about drinking?"

"Fire away, LT."

"When you go ashore after a long, dry, at-sea spell, you slam down a half dozen drinks, and you feel like you haven't started to touch the thirst you built up. That right?"

Twombly nodded agreement.

"When I was enlisted, a chief told me that when I went on liberty, I should limit myself to two drinks. 'Make a pact with yourself that you'll do that,' he said. The first time I tried it, I thought, *Holy crap, I've had two drinks, and I've only been on liberty for fifteen minutes.* But I stuck to the pact. I went back to the chief. He said, 'No power on heaven or earth will be able to stop you from guzzling that first drink. But after you guzzle the first, make that second one last an hour. And when it's gone, go back to the ship. Make a pact with yourself to do that.'"

Twombly said, "Huh."

"You going ashore after we jog?"

"Yes, sir."

"Okay, before you go, make a pact with yourself."

"That an order, sir?"

"No. If it was, it wouldn't do you any good. You understand?"

Twombly said, "Shit."

"Jogging today will be a bit tough, but it'll get easier every day."

Twombly's vocabulary had sunk to a single word.

Teresa X'ed off the date, the last day *Solomons* would be in port in the PI, as Jon called the Philippines. Tomorrow she would X off the first day of the Pacific crossing. She patted her tummy. "Daddy's coming home," she whispered.

The phone rang, and she answered.

"Dearest Teresa."

"Jon! I didn't think you'd call again."

"I just had to hear your voice one more time. The last call was too close to when we lost AB and Skunk."

Jon had paid the phone exchange forty dollars. They'd cut the call off when the money ran out.

Jon gave Jennifer and EJ ninety seconds each, and then he spoke to her.

Teresa had looked forward to receiving a letter from Jon, in which he wrote about the loss of AB and Skunk. She was anxious to see how he was handling it, and she'd gotten no sense at all from their last phone call. He hadn't been able to write, he told her, until last night. After writing, he felt whole again, although he hadn't realized he'd been empty, "If that makes sense," he said.

"Oh, it makes all the sense in the world. I was

worried about how you were handling the loss of your squadron mates, but I didn't know how worried I'd been until I heard your voice and knew I didn't have to be so worried. That makes sense, doesn't it?"

"Teresa Velmer Zachery, if anyone hears our conversation, they will cart us both off to the looney bin."

The one-minute warning came and went. And the parting was sorrow, but it had been dusted with powdered sugar.

She hung up, went to the dining room table, wrote to dearest Jon, and lost her window of opportunity for a morning shower.

Dog Lips was not mad. Dog Lips didn't remember last night, not the snail-eating contest he'd won, the trip back to the ship, or his encounter with Stretch. Stretch visited him in the afternoon and got a list of items to pack for his stay in the hospital and flight home. After his phone call to Teresa, he delivered those. Then he returned to the ship and his and Dog Lips's and Skunk's room. There he wrote to dearest Teresa.

Until taps, which went down at ten. He sealed the envelope. It, like last night's letter, would require extra postage. The PC (postal clerk) was closing out mail at midnight and taking it ashore. He'd be in his cubby hole post office and would grumble when Zachery asked him to weigh his letter.

The PC did frown and grumble. Zachery smiled and paid for the extra stamps. The PC tried not to, but he grinned too, then grumbled, "Damned navy wouldn't be half-bad without dadburned officers."

"Have a good night, PC, and thanks."

"You too, LT, and you're welcome as all get out."

Zachery went back to the room, their room, undressed, and climbed up into Skunk's bunk. There he slept the sleep of an innocent … Little Pootzer.

Zachery woke, hopped down to the deck, did fifty push-ups and fifty sit-ups, and was in and out of the head by reveille at six. It was a habit he'd developed as an enlisted man.

Back then, he berthed in a compartment with thirty-five other men. At reveille, the thirty-five spilled out of their bunks and pushed and shoved to get to the head first. In the head, there were six sinks, two urinals, three commodes, and three showers to service the horde.

"Hurry the hell up," sailors growled in line to use a sink.

"What's taking you so long in there? Eat prunes today, for Christ's sake."

"About goddamned time," as one exited a shower and another shoved in.

As an officer, the commodes-per-person ratio was a more civilized number, but Jon's habit from his days as a seaman had rooted deep. And he liked the feeling he was preparing to face the world while his shipmates still slept.

Zachery ate breakfast, then went to the chapel and

sat in the right rear corner; it was, in a way, his corner. There, a pleasant blanket enveloped him. He knew where the good feeling came from. The two letters he'd written and the last phone call had reestablished his connection to Teresa. He'd often put in his letters that she grounded him, that she bound him to earth, not gravity.

Over the ship's announcing system came, "Underway. Shift colors. The officer of the deck is shifting his watch from the quarterdeck to the bridge."

The ship had taken in all lines securing her to the pier. The flag at the stern was hauled down, one was displayed on the mast, and the function of the officer of the deck, representing the CO of the carrier, had changed from greeting and screening visitors to driving the ship.

Jon said goodbye to the chapel and ascended to the flight deck.

Tugboats had pulled the carrier away from the pier and had begun to spin her around to get her bow pointed to the channel leading out of the harbor. From the starboard side just forward of the island, Jon had a view of Cubi Point Naval Air Station. The runway had obviously been bulldozed out of the jungle. Beyond it and halfway up a hill stood a clutch of buildings hosting the administrative functions of the base, and beyond that area, dense, dark green forest cloaked the top of the high terrain.

The rate at which the tugs spun the carrier picked up. Cubi Point disappeared behind the airplanes parked on the bow. The carrier had visited Subic Bay and Cubi

Point four times during the just-completed deployment. As an ensign, Jon's destroyer had made port visits to the surface navy part of the bay, Subic Bay Naval Station. Now, from his viewpoint, in perhaps forty-five minutes, Subic Bay would have dropped behind USS *Solomons*. Not long after that, the PI would have dropped off the end of the world.

LT Jon Zachery was leaving not only the Philippines, probably forever, but he was leaving the Vietnam War, a war that wasn't won yet.

If I'm not in it, we can't win.

Zachery frowned at his arrogance; then he went below to the room, flopped down the lid to create a desk, wrote dearest Teresa, and waited for her to rise inside him like the sun and shine light and warmth into the dark and chilly corners of his soul. But she didn't rise.

Nevertheless, he made himself write to her.

Teresa poured a cup of coffee for Amy Allison and brought it to her in the living room.

Amy smiled thanks.

Teresa sat on the easy chair with the end table between them.

"Sybil Clark left to stay with her family in San Diego, and now Lauren is leaving too," Amy said. "Before AB and Skunk were shot down, I always thought I'd go home to my family if something happened to Mike. Then I saw how all the wives pulled together to support the two of them, and I thought maybe I'd stay. The people back there wouldn't understand. Here, everybody understands. Now I'm not sure what the right thing to do would be."

"Please, God, you never have to worry about it. But Lauren was thinking about leaving. She told Helen Fischer she'd be a wet blanket on the homecoming. She also said it would hurt to see the other husbands. Then she found out Skunk had been the spare on that last hop, that Dog Lips was scheduled to fly it, but his plane went down. Lauren said if Dog Lips had flown the hop like he was supposed to, he'd have been shot down, and her husband would be on the way home. She did not want to see Dog Lips or be in the same place as him."

"I didn't know that about Dog Lips. Did Jon tell you?"

"No. Helen told me. Sybil called Lauren and told her about Dog Lips being scheduled to fly that last mission, but when his plane developed a problem, they launched Skunk. After Sybil's call, Lauren called Helen."

"But Sybil's in San Diego. How did she find that out?"

"Helen said Sybil's father is a retired admiral."

"So daddy found out about the spare and told Sybil, and she called Lauren. Why would she do that? Even someone as new to navy life as I am knows that would only deepen the wound."

"Helen said she called Sybil and told her that. Sybil hung up on her."

"Well," Amy said, "it has been a blessing to be a part of the Warhorse wives, even with Sybil."

"A blessing for me, too, but I confess I am torn. A part

of me is happy the squadron is decommissioning after the guys get back. Before Jon put in for flight training, we planned on him serving his time and getting out of the service. Then the night of the dog poop happened." Teresa looked to see if Amy remembered. She did. "Jon felt like he had to stay in. He had to serve. He had to fight in the war if so many said they wouldn't. He's done that. Maybe we can—please, God—pick up that old dream and move on. But I will miss you and all of them. Thank you, Amy Allison, for being my neighbor."

Amy smiled. She put her hand on Teresa's arm. "You sound like Mr. Rogers signing off from his TV program."

It was nice to have something to smile, to laugh, about.

Chapter 8

The day USS *Solomons* crossed the international date line, the ship scheduled a memorial service for those lost during the deployment.

The members of the crew not on watch assembled on the flight deck in dress white uniforms and formed up by squadron or ship department.

The Warhorses had been placed in front of the officiating officers' platform. LT Zachery took his place in the front rank. His squadron had, in a sense, the place of honor. To Jon, it was the place of deepest sorrow. The squadron had lost AB and Skunk as recently as an elongated yesterday. And Amos, although Amos had lost himself. He wondered how they would handle LTJG Kane. Maybe they wouldn't even mention him.

The sky was blue with not a whiff of high cirrus to obstruct the midday sun's rays. Many times, Jon had stood in ranks at formal summertime ceremonies, wearing his long-sleeved choker whites, and sweated like a pig, but for the memorial, the ship had turned so that a gentle breeze wafted down the flight deck. The weather was as pleasant as the reason for being there was not.

The formations were called to attention. Two chaplains, CAG, and the carrier CO exited the island and took their places on the platform.

The ship's band played, and a choir from the Protestant community sang the "Star-Spangled Banner" and the first stanza of the "Navy Hymn." The Catholic chaplain delivered an opening prayer.

After him, CAG talked about the demanding profession of naval aviation. The airwing had lost four pilots—half of them in combat but the other two in landing accidents. Naval aviation was as hazardous in peacetime as it was in combat. "Ours is a risky job, but someone has to do it, and thank You, God, our nation keeps producing men who answer the call to duty."

Next the carrier CO spoke. "If you pay attention to the news from home, you can easily get the notion our service is not appreciated, that it is a thing to be spat upon rather than honored. I grew up in a small town in Illinois. We had a Catholic church and a school on a hill. I visited my parents before we departed on cruise. Those good people in my little town are salt of the earth, hardworking, God-fearing men and women, and they appreciate what we are doing.

"I don't know if President Nixon was right when he talked about a silent majority. There are other places in America like the one I grew up in, populated by the same salt-of-the-earth people, and they all honor what we do.

"I believe if you want to get a sense of what America really thinks about our service, you have to listen to the quiet places, not just to the mobs coagulated around a frothing-at-the-mouth yahoo who loves being the center of attention more than he loves his country.

"This deployment is to be the last for *Solomons* and the airwing. The old ship, our old airplanes, and you young men have done your duty well. It has been my honor to serve with you and with our shipmates who gave their lives for our great nation."

After the ceremony, Jon went to the room and wrote to Teresa about the memorial, about how he'd never before felt like applauding after hearing a speech at such a ceremony. He was in ranks and had been ordered to assume the position of parade rest, a notch of relaxation below rigid attention. Military protocol and discipline demanded of him that he remain in that position until his formation leader ordered him out of it. He wrote:

> The CO's speech was extraordinary. For the first time in years, I felt lifted above the protestors back home and, thanks to Amos Kane, here on the ship too. Usually, I feel like they suck me down into the muck they spew. After the CO spoke, I felt elated. I wanted someone to break ranks and clap, but no one did. I asked myself if I am a coward for not being first clapper. Maybe I was.
>
> But it was an extraordinary memorial. Before we left the PI, the ship offloaded a dozen planes, A-4s and F-8s. Air force tankers were lined up to

escort them from the PI to the US, but they arranged it so those twelve planes could fly over the ship at the appropriate moment of the ceremony. As they flew over, single planes peeled out of the formation, each signifying a missing man. It was a gizzard gripper.

After the flyover, the choir sang another stanza of the "Navy Hymn," the "Lord guard and guide the men who fly" one. Then our formations on the flight deck did an about-face, so we could witness a ceremonial burial at sea off the port side while the marines fired a salute.

We did another about-face, and I thought, *Okay, let's wrap this thing up. You've squeezed every drop of juice out of me.* But the Protestant chaplain took the microphone, and he told us the soul of Amos Kane was in that casket along with the souls of the two departed A-4 the two F-8 pilots and that if we wanted closure for the deaths, we had to forgive Amos. If we did that, he said, we'd be able to move on, not forget them but to move on and have a measure of peace with their deaths. The reason we did the ceremony on the date line, he said, was not to leave the memories of our

> shipmates on the far side of the line but rather to carry them across.
>
> After we were dismissed, I went down to the chapel. I sat in my corner, and I made myself whisper, "I forgive you, Amos." But my words were empty of meaning. They were words written in disappearing ink.
>
> Please pray for me, Teresa, that I will be able to say those words and mean them.

He thought about adding, "I feel like my life depends on being able to forgive Amos," but he left it off. It wouldn't have been right to include that thought. It would have made Teresa worry.

Except for the date line crossing, during the transit, Jon's days began, middled out, and ended the same. He woke before reveille, did his push-ups and sit-ups, did his business, ate breakfast, and returned to the room. The day the ship departed the PI, he addressed his first letter of the day to Teresa, but the first letter would never be mailed to her. In it, he wrote about the people he'd killed. As an ensign on a destroyer, he'd killed. He learned how to deal with them in the past, but what happened to AB and Skunk and how it happened, and what happened to Amos and how it happened blew his house of cards coping structure to pieces. He had to start over, to build it anew.

The next day and subsequently, he addressed the first letter to dear God. It was in fact confession. His plan was to place his dear God and the never-to-be-mailed-to-Teresa letters in a weighted bag and throw it over the side before the ship docked in Pearl Harbor.

The second day out of the PI, Tiny knocked on his door at eight. Stretch let him in, and he plopped his hulking frame on the other chair.

"I looked for you yesterday," Tiny said. "Were you in here all day?"

"Pretty much."

"You staying in here today too?"

"Pretty much."

"Anything you want to talk about?"

Stretch didn't look up from his writing. "Nope."

"You want me to just shut up?"

"Pretty much."

Tiny picked up a book from the shelf behind Stretch's bunk and started reading.

At nine fifteen, Stretch said, "I'm going jogging on the flight deck with one of our sailors. You can come if you want."

Jon jogged with Tiny and PO Twombly. Tiny and the PO had a continuous palaver going. Stretch plodded along beside them with his lips zipped. After the run and a shower, Stretch and Tiny ate lunch together, then Tiny returned to his squadron, and Stretch to the room. On the days after the memorial service, he stopped by the chapel first and said an empty, "I forgive you, Amos."

RT showed up at four and sat on the chair. He

brought his own book and read as Stretch wrote. At five, Stretch and he ate a taciturn dinner in the dirty shirt.

RT always invited Stretch to the double-feature movie in the ready room, but he always declined and returned to the room. The evening was when he wrote dearest Teresa letters that went into envelopes and into the mail.

On a Saturday evening, three days before docking at Pearl Harbor, Stretch wrote dear God about the squadron decommissioning, his probable dead-end job, and how he felt he'd failed in discharging the duty he owed his country. He ended it with *Thy will be done*. That evening, he wrote about the same subject to Teresa, and it turned out, he'd written the same letter. He sealed her envelope, dropped it in the slot at the post office, and went to the chapel. From his corner, he forgave Amos in a voice that was pretty much sincere.

Teresa pulled the covers up and whispered, "If I sleep before I pray, and if I die before I wake, I pray to You, dear God, take my soul anyway."

All day, EJ had been whiny, clingy, and demanding. Even Jennifer, normally so eager to be big sister and help with "terrible three" younger brother, also demanded attention and assistance writing her ABCs and one, two, threes.

"You know how to write them, Jennifer."

"No. Show me."

"Show me, please."

"Pleeeeeze!"

Teresa had sighed, placed cranky-pants EJ on her lap, and written a big A and a little a on the paper.

It had been one of those days.

Teresa took her rosary from the nightstand.

Still, thank You, God, I got in a trip to the commissary in the morning and the exchange in the afternoon. Thank You, God, for that, and for Amy.

Her neighbor, Amy Allison, was also three months pregnant. She had flown to Hong Kong with Teresa. That day, Amy had done her commissary shopping with Teresa so she could help with Jennifer and EJ.

Bless Amy, Lord. She is, herself, a blessing to me, but she helps me see my life as filled with Your blessings. Jennifer, EJ, Jon—

Jon.

Three times, she'd started her letter to him, and three times, she'd gotten interrupted. She received letters from him that day. According to Helen Fischer, *Solomons* was able to launch and land a cargo plane to take mail off and to bring airplane spare parts back aboard, but the ship was not able to receive mail. Jon would not get a letter from her until the ship reached Pearl Harbor. Then he would receive fourteen of them.

In their two phone calls from the PI, Jon had sounded so much better on the second one. She'd been hoping she'd sense the same progress in his dealing with the loss of two squadron mates and the treachery of another. In his letters, there was so much of the man she loved since high school, but there was also an element of something else, deep-rooted grief maybe, over …

something more than the loss of AB and Skunk. *Amos Kane and what he did?*

Please, Father God, who art in heaven, touch Jon with the tip of Your healing finger.

EJ rousted her from deep, dreamless, delicious sleep, with his one-word call to duty.

"Breftus."

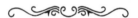

After breakfast on Monday, Jon stopped by the chapel and then returned to the room. He still could not call it his room. Too much of Skunk, too much of Dog Lips abided there. Possession might be nine-tenths of the law, but there was that other tenth. At times, the little one-tenth seemed bigger than his nine.

On the deck in the room, Jon found two envelopes.

One contained a note from Twombly. He was not going to jog with Lieutenant Big Tiny and Lieutenant Little Stretch. That day, he would jog by himself, with the decision to do so all his own idea.

Good man, Twombs.

The other was from Tiny.

Jon sat at the desk, opened it, and pulled out a single sheet of paper.

> Dear Dumb Butt,
>
> Most of us struggle to love our neighbor as ourselves. We're normal. Why the Sam Hill can't you be normal, Jon Zachery? But, no, you have to be

> different. The issue with you is, why can't you love yourself as much as you love your adopted son, PO Twombly?

Tiny had filled the front side of the page with large letters, about the size of the letters Jennifer penciled in a note Teresa included in her letter. Tiny's longhand on the backside shrank to normal size.

> Here's the deal, Dumb Butt. I know how you are about cussing. You think if you let one cussword bounce off your tongue, it will strap you into a toboggan ripping down the chute to dump you into a gaping hole in the top of hell. You are the same way with booze. Two drinks, okay. Three drinks, toboggan to hell.
>
> Something about the last mission we flew has you at the top of the chute, and you are wrestling with yourself as to whether you need to step in and release the brake. All I can think of is that it must have something to do with Amos goddamned Kane.
>
> Dumb Butt, if you go to hell, it's St. Pete's job to send you there. You don't have to do his job for him. He's got nothing else to do anyway.
>
> So, besides being dumb, you are stubborn as a Missouri mule, but see

if you can just do this for me. Let this notion into your box-of-rocks brain: what you did for Twombly was a good work, and it transformed him from being a good guy with a drinking problem to a good man. Period. No qualifiers. And let in another notion. This one goes: "I'm a better man than I thought I was." I know it's too much for you to go all the way to "I'm a good man."

So anyway, today, I got stuff to do. Tomorrow we pull into Pearl. You'll call Teresa. Please tell her, her favorite Gorilla says hey. The second night in port, let's go to dinner in town. My treat. I know what a cheapskate you are.

We got a date?

Jon leaned back in his chair and closed his eyes. He felt … peaceful if not peace filled. It was, he thought, like waking one morning and wondering why he felt so good and then realizing the headache he'd carried around for a week was gone.

An image formed. A gorilla wearing a Lucy mask says, "You're a good man, Charlie Brown."

Jon snorted. "Tiny!"

He opened his eyes, opened his safe, and removed all the dearest Teresa letters that would never be delivered and the dear God ones that had been delivered in the act of writing them; then he placed them in a plastic bag.

His father popped to mind. Saturday. The day before

classes let out for the summer after he completed seventh grade. Pop had rousted him out of bed, told him to eat breakfast, and then to put his bike in the trunk of the Plymouth. Pop then drove him out to a farm. There, he told Jon he was working for Mr. Hemsath.

Hemsath took him to the milk barn and told him how to clean out the cow poop. Then the farmer went inside to get his breakfast. Jon worked for the man for two weeks. No mention had been made of pay. Jon asked.

"Your pop said I didn't have to pay you," Hemsath replied.

That night, Jon asked Pop.

"A boy needs to learn how to work. He don't need no practice learning how to get paid."

Jon shook his head. *Thirty years old, and I finally see you're not dumb and a stick in the mud. Sorry, Pop.*

He poked holes in the bag, carried it to the fantail, and tossed it into the carrier's wake. The bag bobbed and swirled in the eddies, sank and rose again; then it sank for good.

Amos dadburned Kane, I forgive you. And Tuesday and Botch. And, yeah, myself.

Then he walked back to his room, but he stopped by the chapel first.

PART 2
AN EPHEMERON OF ABSOLUTION

Chapter 9

Today's date wasn't circled. Instead, Teresa had smeared on red lipstick and kissed the box on the calendar. After the fly-in, after the initial passion-filled but appropriately restrained greeting in front of the hangar, they'd driven home and entered the house from the carport. He saw the date and kissed it.

"Rats," he'd said. "You can't French kiss a calendar."

She expected him to administer one to her, but Jennifer asked, "What's a French kiss, Daddy?"

The children looked up at him.

"I'll tell you on your thirtieth birthday."

"You have to be careful about what you say in front of Little Miss Big Ears," Teresa said.

"Daddy," Jennifer said, "I can write my ABCs and one, two, threes. I'll show you."

EJ grabbed Jon's hand. "I show you my trucks."

"Here's what we'll do," Jon said. "Jennifer, get out your paper and a pencil while EJ shows me a truck. Then you show me how to write ABCs. Then he can show me another truck, and after that, you can write one, two, threes."

Teresa expected Jon to spend a few minutes with each child and then to encourage them to practice writing and playing with the trucks while he took her to the bedroom, but he didn't. After witnessing his daughter's facility with a pencil and the admirable characteristics of a fire truck with ladders, he sat on the sofa and, with a child on either side, read stories to them.

After lunch, Jon took the children for a walk. Jennifer wanted to see the school she'd attend in the fall, *Kiddy Garden*. When they got back, Teresa was in the kitchen studying a cookbook on the counter.

"We're having beef strogh enough for dinner," Jennifer said. "Your favorite."

"Yes, it is. I like it lots better than beef strogh not enough. That would mean there wouldn't be any for me."

"Daddy!"

Jon looked at Teresa. He didn't have to say a word. She knew they shared the thought: *So grown-up. Already!*

EJ yawned.

"Sometimes EJ takes a nap," Jennifer said. "I'm too grown-up for a nap."

Jon said, "Rats!"

Teresa'd begun to think he wasn't interested in her, in making love to her. She'd expected him to find a way to get the children occupied. She put her hand on Jon's arm, and he looked into her eyes with such longing and hunger and need.

She took Jon's hand, and they led EJ to the children's room and put him to bed. Then Teresa pulled Jon back to the dining room table where Jennifer was coloring.

"Daddy and Mommy have to talk, Jennifer," Teresa

said. "See the clock? The big hand is on nine. We'll be done talking when it gets to three. Okay?"

Jennifer nodded and went back to her coloring.

In the bedroom, they sat on Teresa's side of the bed. She started unbuttoning her blouse. Jon sat and stared at the wall.

"What is it, Jon?"

"I'm afraid I'll hurt you. And Little Pootzer."

"Jon, we've talked about Daniel. We've written about him dozens of times. We did make love the night we lost him, but that isn't what caused him to come early. I thought you accepted that. Besides, I talked to my doctor. He said making love is okay as long as we don't get into bodice-ripping, wild and sweaty banging of bodies."

"But that's what I want to do to ... with you."

Jon thought, *And once we get started, I'm afraid I won't be able to hold back. Just like I couldn't keep myself from recharging my guns.*

The deaths on his soul, he knew he hadn't put them in the plastic bag, along with his letters to God, and dropped them off the fantail. Those deaths would be with him forever. But everyone he killed had intended to kill his squadron mates or shipmates. It wasn't just his own life in consideration. And it had been in balance. For a week almost.

But now, sitting on a bed next to the most desirable woman in the world, her blouse open and her bra visible, lust and death swirled inside him like a tornado ripping up Kansas. The want in him was a cauldron about to boil over. Wild and sweaty banging of bodies was right there, and all it needed for the release was a touch, a look, a kiss, the undoing of one more button. After the wild sex, there'd be a trip to the hospital and the loss of Little Pootzer, just like with Daniel.

"Jon Zachery," Teresa said. "We are going to make love, and this is how we are going to do it. I will make love to you, and you will take it like a man."

Which he managed to do. Afterward, they lay with their arms around each other and their legs entwined. The only thing wrong was there was no way to get enough of himself in contact with enough of her.

"Teresa."

"Mmmmm."

He squeezed her a little tighter. "I discovered something about myself crossing the pond. I always thought it was important that I be guided by one morality that applied at work and at home. I looked at some of the other guys, and I thought they came to work, hung their uniform and at-home morality on hooks in their locker, pulled out at-work moralities and flight suits, and went to work. I looked down my moral nose at them. Those two weeks between the PI and Pearl, I thought about AB and Skunk and Amos a lot, plus a few other things. I concluded I have no business looking down my moral nose at anyone. I'm just like the rest of them. Right now, my at-work morality is in the bag with my flight gear in the trunk of the car. I haven't figured out how to deal with this way of looking at myself."

Teresa didn't say anything.

That was stupid, stupid, stupid. You should've left that in the box of things you don't tell her.

"Helen Fischer and I talk," Teresa said. "She says RT thinks you see the black-and-white of issues more clearly than anyone he knows, and then you act accordingly. RT thinks it's your most admirable quality."

That's what I think of RT.

"This is what I think, Jon Zachery. There is one set of moral principles guiding you. It's just that at work and at home, that set of principles has to deal with different circumstances."

"Huh." He lifted up on an elbow. Mickey's big hand on the clock was on two.

"I'm going to tell Jennifer we're not finished talking."

"That's because you didn't say a word until a minute ago."

"Well, it wasn't my fault."

"Come on. I have to get dinner started."

"I meant to tell you. I have a new favorite dinner."

"And that is?"

"Peanut butter and jelly sandwiches."

She pushed away from him. "Have you heard of deferred gratification?"

"Yes, and it stinks to high heaven."

She closed her robe, tied the belt, and smiled. She was happy he saw and thought, as he did so often, how she looked like the *Pieta*. Except the Pieta's face was laden with sorrow as she revisited the suffering and awful death her child endured. But Teresa was, at that moment, happy. Maybe she saw love and life ahead for Little Pootzer.

Please, God.

After beef stroganoff and after the children were bathed and put in pajamas, Jon washed dishes as Teresa dried.

Through the window above the sink, he saw across adjoining backyards to the Allisons' kitchen window.

"Huh," Jon said. "The window's dark. I bet they ate dinner and left the dishes until tomorrow."

"Keep washing, buster."

"Yes, dear."

He rinsed a plate and placed it in the drying rack.

"Teresa." It was, he thought, a serious-sounding Teresa.

She looked at him and elevated her left eyebrow. He shook his head and couldn't keep the smile from breaking apart the serious face he'd put on. She could levitate her eyebrows independently. He'd practiced in front of the bathroom mirror, but his facial muscles had been unable to match Teresa's ability. She liked to tease him about his lack of that special talent.

"Yes, dear?" she invited.

"I know you were happy to think we were decommissioning, that after two years in a dead-end job, I'd leave the service. I spent a good bit of the trip across the pond trying to reconcile myself to what was going to happen to me. Not to us but to me. The notion that what I'd seen as my duty to serve wasn't worth a hoot to the navy. It was like my one morality thing. What I saw as my duty had been like bedrock. Who I was stood on that. And the bedrock crumbled. Finally, something Pop said to me settled it all for me. It was kind of like, 'Work hard, take care of what you can take care of, and let the Lord handle the rest.'"

"Yes, dear?" An invitation to spill the part he hadn't yet voiced.

"It was the day before we pulled into Pearl when I thought of what Pop told me. You've told me, 'If you have something troubling you, let go of it. Just give it to the Lord.' That letting go of a trouble was something I'd never been able to do. If something troubled me, it had to be something important. I always felt like if I let it go, I'd lose an essential part of me. But that day, I let it go. And I felt so good, so free of ... burden."

"When you called from Hawaii, I was so pleased to hear not just your voice but what it contained. I worried you across the Pacific. I worried about how you were handling what happened on that last hop. I worried about how you were going to handle the decommissioning, but you sounded so at peace with everything."

"I was. For all of an hour. When I got back to the ship, Tiny was camped outside my room with the news we were going back to Nam one more time. Tiny was majorly fired up. I don't know what I was."

"Wash," Teresa said.

"Oh." He recommenced scrubbing out caked-on goop from a saucepan. "I had this worry. I let go of it and gave it to the Lord. All of a sudden, I could handle it. It no longer tried to eat a hole in my belly. Then, the worry I had was no longer there. I was not going to a dead-end job. Then I had a new worry. You were happy with the notion of me not going back to the war. I told God, 'If to make me happy, You make Teresa sad, I don't want to be happy.'"

Jon rinsed the final pan, placed it on the rack, and let the water out of the sink. "There," he said. "Today's mess all cleaned up so all we have to do tomorrow is face tomorrow's messes, right, Teresa Velmer Zachery?"

She dried the pan and handed the dish towel to Jon. "Would you like to use the bathroom first?"

"Ladies before gentlemen."

She levitated her left eyebrow. It meant, *Is there one of those around?*

When it was his turn, he decided to try simultaneous brushing of teeth and shaving. In the end, it cost him time instead of saving it; plus, he nicked himself twice. Dealing with that cost him more time.

He expected to find Teresa in bed in a nightgown. Instead, she sat *on* the bed, fully clothed.

She stood and kissed a fingertip and touched it to his wounds. Then she handed him his rosary and invited him to kneel beside her.

She tugged on his hand until he knelt.

Teresa used her "Thy will be done" mysteries of the rosary. In 1968, she'd found the church-sanctioned mysteries inadequate to her needs. The navy had sent them to Meridian, Mississippi, where Jon entered training to fly jets. On their first Sunday in town, they went to Mass and found the church jammed with people, except for one pew that held only a single young colored girl. It turned out she was sitting in on the service in the white church. Their usher tried to shoehorn them into one of the crowded pews, but Jon pushed him aside, and they entered the pew with the girl. People stared and radiated chilly disapproval at them, but otherwise, no one bothered them—until that night when some Klansmen tacked a threatening note to their front door. Over the next weeks, there were other reminders from the Klan that the Zacherys were being watched. Teresa developed the "Terrorful Mysteries" to get them through their Meridian tour of duty.

Teresa said, "The second Thy will be done mystery. Your ways, oh Lord, are not known to me, a sinner—"

"To us sinners."

"To us sinners, Lord. Thy will be done."

Jon thought he might create his own mysteries of the rosary. He'd call them the deferred gratification mysteries.

Teresa hadn't started the Our Father, which should have come on the heels of the second mystery. Jon looked at her. She pulled him to his feet and unbuttoned the top button of his shirt.

"Uh, Teresa, can you read my mind?"

"Like a book." Then she undid a second button.

Later, she lay on her back, Jon on his side with his hand on Little Pootzer. She said, "When I found out you were going back to war again, I was mad at God. I told Him, 'You hold out to me the end of my husband having to fight, and I reach for it, and You snatch it away.' I was so mad. But I didn't stop praying, and that saved even a wretch like me. In the prayers, I found reinforcement of my belief that God will not give us a challenge we cannot handle."

"Do you know what I like to do with you, Teresa?"

She raised up on an elbow, and he flopped onto his back. The nightlight cast just enough illumination to see that the left eyebrow had elevated.

"I like to talk with you."

The left eyebrow fell down. She giggled; then she laughed.

"Shhh!"

She kissed him, snuggled against him, draped her arm across his chest, and sighed out the cares of every person on earth. He thought she was slipping into sleep, but she giggled again, then laughed again. Her belly jiggled. It was as if Little Pootzer laughed at him too.

Who'd have thought being laughed at would be so close to heaven on earth?

The next day, the *Solomons* docked at Alameda Naval Air Station on San Francisco Bay. The crew boarded four navy transport planes and were flown to Lemoore. Twombly had asked to be introduced to Jon's family.

Jon drove them out to the airfield in time to see the transports land.

Out of an ocean of sailors, Twombly found them in a sea of families. Teresa thought he was genuinely happy to meet the lieutenant's family, and she was impressed with him. After some twenty minutes, Twombly boarded a bus for the barracks, and the Zacherys drove home.

"He's a good kid," Jon said. "Out of a couple thousand, he's one of the best."

"Petty Officer Twombly thinks Lieutenant Zachery is the best."

"Petty Officer Twombly is *the* best," Jennifer said. "He gave EJ and me a Tootsie Pop."

Jon said, "If I give you a Tootsie Pop, will I be one of the best?"

"Yes."

"It'll take two of them to butter me up, buster," Teresa said.

Except she would never eat a Tootsie Pop. She would worry it'd make her cheeks puffy.

As he drove, she squeezed Jon's arm to her. Out the window on his side, the Sierra Mountains walled in that side of the San Joaquin Valley. The Coastal Range did the same out her window. Ahead, but not visible in the haze, the mountains separated the valley from LA. The San Joaquin Valley was a huge basin. It was almost big enough, she thought, to contain the happiness in her heart.

Jon had fourteen days of leave. Two of them had been spent. At the end of them, Jon would have to go back to work. And it would be busy for him. The squadron would have to cram a full year's worth of training into half the time, and that with so many new pilots and so many new enlisted men to maintain their planes.

But they had twelve more days, and she was

determined to fill each precious second with as much love and life as it could hold.

Tomorrow, Jon would get his cruise box from the ship. After a cruise, his clothes always smelled like cigarette smoke. His stinky stuff would go directly into the washing machine and get rid, in a way, of yesterday's mess. On Friday, they would drive to Los Angeles for a visit with EJ's godparents, civilians they'd met while Jon was in college. After the weekend with them, they'd drive along the coast to Monterrey and spend two nights there. Then they would come back to their clean and clean-smelling home and still have a few more days of this time that was even better than Hong Kong had been.

Chapter 10

Jon said, "EJ, eat your cereal."

"We bisit Pope Unca Herman." The boy's face glowed as if his halo had melted.

During college, the Silings became friends with the Zacherys and godparents to EJ. Jon's navy friends at Purdue dubbed religious Herman "Pope Herman." The moniker stuck like a pilot call sign. Unable to have children of their own, Pope and Mrs. Pope, Ethyl, doted on Jennifer and EJ. EJ was particularly fond of being doted on.

"Drink your juice." Jon tapped his finger by the glass.

"We go to Dizzy Land."

"Yes. Today we drive to Los Angeles and see Pope Unca Herman and Mrs. Pope. Tomorrow we go to Dizzy Land, but we are not going to leave until you eat your breakfast."

That message sank in. Too well.

"EJ, slow down. Chew your food."

The phone rang. Seven o'clock. *This can't be good news.* Teresa answered the wall phone in the kitchen, listened, and held it out toward him. "Jon. It's the new skipper."

He'd heard the Bureau of Personnel tabbed someone to be the new CO, but he didn't know the man's name. "Lieutenant Zachery, sir."

"I am not the new skipper; I am *the* skipper."

"I thought there was to be a change of command next week. Sir."

"We just had it in the admiral's office. Now there's no time to fumble-fart around here. I need you in my office in fifteen minutes."

"I'm on leave, sir."

"Your leave is canceled. My office. In fourteen minutes."

"It'll take thirty-five minutes to get out there, sir."

"Thirteen minutes."

"Commander—"

"Call me Skipper."

"Sir, if I don't get stopped for speeding, I can make it in fifteen. But I'll be in ratty civvies and no shave. If I show up like that, are you going to send me back home to get properly dressed?"

"Zachery, you're pissing me off."

"Sorry, sir. See you in fifteen."

Jon hung up and turned.

"Your leave is canceled. I didn't have any trouble hearing that." More than her words, the look on Teresa's face hit him as if he'd been running full tilt in total darkness and smacked into a brick wall. When Daniel died, he hadn't been in the hospital with her. He'd been there for the midnight emergency C-section, the interminable wait for Teresa to come out from under the anesthesia, and for more hours as they held hands and prayed for their preemie. At eight in the morning, Teresa's doctor told Jon, "Your wife is doing well. Your son is hanging on. Go home. Get a shower; then come back."

Daniel died when he was in the shower. Teresa took the news alone, without him there to share some of the load. But now he was sure he knew the look that would have lived on her face.

He left with Jennifer crying, EJ howling, and Teresa trying her best to console them with no time left over to deal with what she lost.

In the carport, he said, "Hello, car." When he flew, "Hello, airplane," put all his concerns in a trunk in the corner of the attic of his mind so 98 percent of his mind had nothing in it but flying—or, in this case, driving.

He drove five miles over the speed limit through the family housing area, through the base admin area, speed limit plus ten, and on the road through farmland to the airfield, plus twenty.

Jon busted into the squadron admin office at 7:19. Fifteen minutes give or take.

The CO's office door was closed. Petty Officer First Class Wazinski looked up from behind his desk.

"PO1," Jon said. "Our new CO, what's his name?"

"Commander Fuller, sir. Uh"—he lowered his voice—"he's an East Coast puke."

There were two Iron Curtains. The second one sat over the Mississippi River and segregated East and West Coast segments of the US Navy. West Coast sailors talked about those on the other side as East Coast pukes. The word flew over the wall from the other side as well.

Waz knocked on the CO's door and opened it. "Lieutenant Zachery's here, sir."

"Get your ass in here, Lieutenant."

Waz stood aside. Jon entered and started closing the door.

"Leave it open."

His new CO sat behind his desk. He was skinny with close-cropped black hair. The man's dark eyes blazed with missionary zeal. Those eyes swept Zachery up and down. He didn't say anything.

What kind of stupid game is he playing?

The CO stood. "Follow me." He brushed past Jon, out through admin, and into the passageway. Zachery had to trot to catch up.

Outside the hangar, the CO strode to a two-door Mercedes. "Get in."

Zachery sat on a leather seat in a classy vehicle that smelled new.

Fuller said, "You don't have a poker face, *Lieutenant*." Disdain dripped off *Lieutenant*. "Your attitude is a problem. We're going to drive for five minutes. When we get to where we're going, you best"—he used the gerund form of the F-word—"have your attitude squared away."

Jon gritted his teeth. The pompous prick had invaded his happy house and left it filled with the sounds and sight of heartbreak.

The new CO's uniform looked tailor made. For that matter, he looked tailor made. He backed out of his parking space. "Another thing. I read JOs' minds."

A TICKET TO HELL: ON OTHER MEN'S SINS

Jon thought of his tour of duty on a destroyer as an ensign. His first CO had been a great skipper, as had been the case in the Warhorses. The second CO on the destroyer had been a horse's behind, but Ensign Zachery had learned to manage his attitude.

Lieutenant Zachery snuck a glance at his second Warhorse CO, to see if the man really could read his thoughts. What he saw surprised him.

Both CDR Fuller and RT had an *eternal boy* face. They would both lie in their dead-of-old-age coffins looking like an unwrinkled seventeen-year-old. In the heat lightning flashing through their first meeting, it hadn't registered. Now it did. One difference though: RT projected a mien of *If there's anything I can do to help you, let me know.* Fuller was a lightweight alpha male. He'd never be able to physically cow a challenger by puffing out his chest and snarling. Rather, during the posturing phase, he'd go right for the throat and leave wannabe alpha bleeding out in front of the pack.

Jon wondered if his children had stopped crying. He thought about what Teresa said about God not giving them a challenge they couldn't handle. He tried to form the words: *Thy will be done.* But it was as if those words formed on a neon sign, only the sign wouldn't stay lit.

The CO drove beneath the underpass supporting the ground-level taxiway and turned right by the tower, which stuck up into the sky like the Washington Monument. A city block later, the CO pulled up and parked outside a long, high-sided, single-story, concrete block structure with a fence around it.

There was no sign on the fence or the building, but

Zachery knew it to be the Weapons Training Facility. Both pilots and ordnancemen received training there on the more sophisticated weapons the A-4 could carry and deliver. Much of the material covered inside the drab structure was classified.

Fuller exited the car. "Zachery, you asleep in there?"

Thy will be done.

The commander punched a code into a keypad on the side of the gate through the fence. Normally, the gate was open during workdays. Another code opened the door to the building. Inside, a long corridor led the length of the structure. To the left, Zachery knew, were classrooms. To the right, two large bays held mockups of A-4 and A-7 aircraft where ordnancemen trained loading weapons.

Fuller pulled a plastic-covered card from a shirt pocket, clipped it on, and led them to the end of the corridor where a US Marine sat behind a desk. Jon had attended top secret briefings in the building, but there'd been no marine guard for those.

The marine studied Fuller's badge. Then he shifted his gaze.

"Give him your ID card," the CO said.

Jon did, and the marine told him he'd get it back when he left.

Strange. Never had to go through this kind of security drill before.

The sentry punched a button on his desk.

Fuller walked around the desk and opened a door into a room. As the door opened, an alarm bell rang from inside.

"Zachery, get your ass inside so I can shut the damn door and turn off the damned bell."

A lieutenant commander jumped to his feet from behind a table and grinned. "Great to see you again, Commander, and congrats on the command."

The LCDR was tall, bald on top, close-trimmed black hair above his ears, thick black moustache. Pale complexion. Didn't get out much.

"Good to see you, Will." Fuller pulled out a chair from behind a table and sat.

The room was a standard navy classroom—metal tables, metal folding chairs, a screen pulled down out of the ceiling.

Doesn't-get-out-much sat by a projector with a stack of viewgraphs next to his elbow. He said, "Will Morrison, Lieutenant Zachery. CDR Fuller and I served in the Pentagon together. We just escaped from the Puzzle Palace."

"Let's get on with this, Will. I have an appointment with the admiral in forty-five minutes."

"Yes, sir." Morrison switched his attention. "Take a seat there." The LCDR pointed to a table. "Where do you think you are, Lieutenant?"

"If there's some classification level higher than top secret, this is it, sir."

"Maybe I'll clarify this for you later," Morrison said.

Fuller piped in. "I need a man for a special job. Your previous CO said I should pick you. We'll see. Tell us what you know about SAMs as employed by the North Vietnamese."

Zachery took a moment to marshal his thoughts,

sort of like getting them into formation and making sure the rows were straight.

"First thing, sir, surface-to-air missiles are part of an overall air defense network consisting of long-range air search radars, MiG fighters, AAA, SAMs, and comms to tie them all together."

"Zachery," Fuller growled, "I asked you what time it is, not how to build a goddamned watch factory."

Jon turned. "Commander—"

"I told you to call me Skipper."

"Yes you did, sir, and, sir, I am trying to tell you if you want to understand the threat posed by SAMs, you have to understand the North Viets do not employ them as isolated entities. They are part of a network, and the network makes the SAM more formidable than if it was isolated."

"He's right, Commander Fuller."

Fuller glared at Morrison.

Morrison dropped his gaze to his hands on the table, for a moment. "He *is* answering your question, Commander. Go ahead, Lieutenant."

Zachery described the Guideline Missile and the Fansong target-tracking radar. "I don't know much about the control van, but I do know the North Viet SAM operators use inputs from the overall air defense network to minimize warning time to us in our cockpits."

Morrison said, "How do you know all this, Lieutenant?"

"From LCDR Fischer, sir. He's the maintenance officer in our squadron. Everything I know about it came from him."

A TICKET TO HELL: ON OTHER MEN'S SINS

Fuller humphed. "Fischer. He's up for commander with fifteen years in, and he's resigning his commission."

Fuller glared at Jon. Jon did what Morrison did. He looked down for a moment, long enough to forfeit the stare down, then up again.

"Commander Fuller," Morrison said, "if you want to accomplish your goal, Zachery is an asset you're going to need."

"Fine. Brief him into the project."

Fuller left. When the door closed behind him, the alarm bell extinguished and left behind a deep and profound silence. It would take some looking to find God in it though.

The LCDR said, "Commander Fuller is smart, ambitious, and connected. He came up with an idea to improve how carriers fight in Vietnam. Each of the A-4 squadrons will specialize in a particular mission. One will master laser-guided bombs, another the TV-guided Walleye weapon, and the Warhorses will be SAM fighters. To get the navy bureaucracy to buy into this concept, it took all his strength and determination. I don't think any other commander could have gotten it done."

He's also as dangerous and cranky as a rattlesnake with its rattles chopped off.

Morrison slid a sheet of paper across the table. Lieutenant Jon Zachery was to be briefed into Project Little Round Top. The project was top secret, need-to-know, and compartmented. Compartmented meant he could not discuss it with anyone not briefed into Little Round Top. And if they were discussing the project and the alarm bell rang to signal someone was entering the room, they had to cease talking until the visitor closed the door behind him *and* they verified that the visitor was briefed into Little Round Top.

The project concerned a complete SAM system the US had acquired. Jon did not need to know how, who from, or any of those kinds of details and was, in fact, forbidden from asking such questions of even others briefed into the project.

"Tomorrow, Zachery, you're going to fly to Nellis Air Force Base. The—"

"Tomorrow! I'm on leave."

"CDR Fuller told me he canceled your leave." Morrison shrugged.

"So, the SAM system is near Nellis. Air force intel types will brief you and then take you to tour the system."

"Tomorrow?"

"Yes, and Monday you'll fly to China Lake, where you'll get a briefing on a new SAM fighting system the navy developed. You'll spend three days there. The system, by the way, is called *tie us*, spelled T I A S. The letters stand for target identification and acquisition system. Your new CO, by the way, would have had you fly over on Sunday, but China Lake doesn't work on weekends, unless you're willing to pay a lot of money for overtime. And our TIAS budget doesn't have the money for that."

"Judas priest!"

"You want sympathy, Zachery? It's—"

"I know where it is in the dictionary. Sir."

A TICKET TO HELL: ON OTHER MEN'S SINS

Teresa said, "Jon, you have to eat something. You're flying tomorrow."

During Jon's first stage of flight training, a chaplain had spoken to the wives of the student naval aviators. His message had been, "Your husbands are engaged in a challenging"—Teresa and the other wives had known the man had intentionally avoided the word *dangerous*—"occupation. They will learn how to handle the challenges if they keep their minds on their business. You can help them by minimizing stress in your relationship at home." Teresa had retained the message as "If you are not careful, you will kill your husband."

The chaplain's message poked at her now.

EJ had cleaned up his plate. Jon reached for it to give him more.

"No! Mommy do it."

"EJ, let your father help you."

"No!"

The anguish crossing Jon's face gripped Teresa's heart with clawed fingers.

"Daddy needs a hug." Jennifer hopped down from her chair and administered one.

Teresa walked around the table, around the little hurt-filled boy in his highchair, to hug Jon's other side.

Then she and her daughter returned to their places and forked up a bite from their plates. Jon, too, ate. Then her eyes met his. Tears glistered their images of each other, or maybe she was the only one seeing through overmoistened eyes.

Chapter 11

Stretch taxied past the tower and the admiral's building and parked in the holding area short of the takeoff end of runway 32. Not a cloud in the sky, visibility clear to the moon. He ran through the takeoff checklist and a review of takeoff emergency procedures.

Ready to go flying, 510? Speak up if you aren't.

Stretch keyed the radio, "Tower 510 for takeoff."

"Five ten, cleared for takeoff. Switch to departure control on 313.4."

He taxied into position, snapped his oxygen mask in place, and gave himself a last piece of advice. *Head out of your butt.*

He set the throttle to 90 percent of full power and checked the engine gauges as his right hand moved the stick around the cockpit to ensure there was no binding in the flight controls. All good. Feet off the brakes, throttle to full.

Light braking kept the jet lined up with the runway. The rudder became effective. Stretch's eyes dropped to the gauges on the instrument panel. All good, but his eyes moved too slowly. Pilots talked about being "behind the jet." And that he was. It felt as his plane moved faster and faster, he could not keep up with it, as if it were getting away from him.

Judas priest!

Ten knots above the speed where he should have pulled back on the stick, he pulled, and 510 leapt into the air, a demented demon just busted loose from hell rather than as it should have been—smoothly slipping from weight on the wheels to weight on the wings like a ballerina supported by a male partner rising ever so gracefully from the stage to floating above it.

Landing gear up. No! Not the flaps yet.

Now the flaps. Raise the flaps too soon, and the jet could stall.

Clunk, clunk. The wheels snugged into the wheel wells. And like that, he was flying with the airplane, no longer trying to catch up.

"Five ten, departure, turn right to one zero zero."

"Five ten to one zero zero."

As he turned to the right, he thought, *Saying, "Hello, airplane," isn't a magic formula that turns off other-than-flying concerns. What turns off the concerns is saying the words with conviction. And commitment.*

He rolled out of his turn, flying directly toward the eyeball-frying sun. If another plane was coming at him, he would not see it. At times, there was faith involved in flying, and a pilot relied on the *It's a big sky, and pilots hardly ever run into each other* philosophy. Faith, he thought, got a pilot through flying directly into the sun. It also got a pilot through periods when his head was other places than on straight.

But he was through that now. Stretch and side number 510, through a spiritual symbiosis, morphed into a single entity, empowered to be like the line from the poem by that US Air Force guy, *soar above the surly bonds of earth*. To soar while feelings of immortality and superpower pulsed through the veins of the creature of aluminum and flesh and hydraulic fluid and blood.

At that moment, to be a pilot cut free of gravity's *surly* bonds that tied earthbound mortals to the planet, to *zorch* through the sky at four hundred miles an hour … well, it was a glorious feeling.

Some pilots described flying as being the most fun a guy could have with his clothes on, but this was not a feeling to be sullied with crudity.

Thank You, God, for wings.

"Five ten, switch to LA Center."

After landing and parking, Jon entered the Nellis Passenger Terminal / Flight Operations Center.

"Lieutenant Zachery."

An air force second lieutenant rose from a seat. Under a flattop, the kid had Oriental features on the smooth-skinned face of a nineteen-year-old wearing an I'm-all-business demeanor.

"Show me your ID card," All Business said.

Zachery frowned at the kid. "Show me yours."

All Business drew back.

"Show me your dadburned ID card, Second Lieutenant."

The kid produced it. Second Lieutenant Leo Hu.

"*Hoo*. Did I pronounce that right, Lieutenant? And you're in intelligence, right?"

The kid's Adam's apple bobbed. "Yes, sir, that's right. Uh. On both counts, sir."

"I'm pleased to meet you, Lieutenant. Here's my ID. I'm Jon, by the way, without an H. My pop didn't have much schooling, and Mom had him fill out the paperwork for my birth certificate. After me, she did the paperwork for my brothers and sister. Isn't that interesting?"

"Uh, yes, sir." He returned Jon's ID.

"One more question, Lieutenant Hu. Is everybody I meet today going to treat me the same way you greeted me?"

"Sorry, sir. Um, can we go now, sir?"

Outside the terminal, they made their way to a red Corvette with the top down. Lieutenant Hu exited the parking lot and drove past a "Speed Limit 30" sign at precisely 30 MPH. The navy lieutenant looked at the air force one. The kid, he was sure, was rebuilding his intel-wienie-I-know-stuff-you-don't-know façade brick by brick.

The second louie parked against a curb in a reserved parking spot next to a sidewalk leading through a lawn of desert sand to a drab box of a building. Warhorse sailors described Lemoore Naval Air Station as the middle of—gerund form of the F word—nowhere. From what Zachery had seen of it, Nellis AFB was even more nowhere than Lemoore.

Lieutenant Hu led Jon thought a gate, through a

door, both with keypads, and into a corridor very much like the one at the Weapons Training Facility. At the far end, a security man sat at a desk. The procedure Jon had to pass through was identical to the one at Lemoore.

Inside a briefing room, a man waited, a Caucasian and old. Forty-something probably. Hu was slender, this guy hefty with saggy jowls, and his crew cut was grizzle gilded. His expression was the same and easier to interpret. Forty-Something did not want to be where he was on a Saturday morning.

Jon stuck his hand out. "Jon Zachery, lieutenant, United States Navy. I'll bet you a hundred bucks I got a bigger reason than you to *not* want to be here on a Saturday morning."

"You just got back from an eight-month deployment. No bet, Navy. I'm Fred." Fred shook the proffered hand. "This morning, I think we can tell you some stuff you'll find worthwhile. This afternoon, we'll show you something that will definitely be worthwhile."

The briefing room was similar to the one at the Weapons Training Facility, except the desks and chairs looked newer. Or maybe sailors abused furniture more than air force guys did. The bulkheads, walls the air force called them, were adorned with large photos of desert scenes: a cactus sporting a red blossom, a long-tailed mouse, a coyote, a sidewinder with wavy lines left in the sand behind it.

Fred got Jon a cup of coffee from a pot on a sidebar while LT Hu began flipping viewgraphs. Hu covered the air defense system employed by the North Vietnamese in more detail than LCDR Morrison had yesterday.

Fred gave the second brief, which covered MiGs, North Vietnamese employment tactics and the strengths and weaknesses of each model they operated.

The next briefing concerned the SA-2 Guideline Missile, the Fansong target-tracking radar, and how the North Vietnamese used the overall air defense system to maximize the effectiveness of their system while minimizing the threat from US countermeasures—subjects that had been covered by Morrison yesterday. But he had constructed a skeleton. These air force intel types had hung the meat of detail on the bones.

Zachery told Fred, "I knew some of this stuff at the surface level, but I sure didn't know the details. It was worth coming over on a Saturday to hear this. Thanks."

Fred nodded a *You're welcome, Navy.*

LT Hu put a viewgraph on the projector. His lecture concerned how the Viets employed the Fansong radar to minimize its exposure to Shrike missiles fired from US Air Force and US Navy planes.

"As employed by North Vietnam, the Guideline missile will fly to intercept a target plane in about forty seconds. The Fansong radar will be on the air for thirty to thirty-five seconds. The flight time of a Shrike you fire at the Fansong will be in the neighborhood of ninety seconds. The obvious conclusion is that you don't want to use Shrikes to fight the SAMs. Find the sites and bomb them the way US Air Force Wild Weasels do."

"I thought Weasels carry Shrikes," Jon said.

"Yes, they do," Fred piped in. "But they also carry bombs, and before the bombing halt, they killed more SAM sites with bombs than with Shrikes. But one thing

Leo left out was that the North Viets respect the Shrike missile. That's why they minimize the time they keep the Fansong radar on the air."

Hu jumped back in. "But the missile flight time is not the only issue with Shrikes. Before you fire one, you have to pinpoint the location of the SAM site, then pull the nose of your aircraft up to shoot. The whole process can take a minute. While you are diddle-dorking around pinpointing the location, a missile is in the air and tracking you. You are dead twenty seconds before you even get a Shrike off. Even a dipshit navy pilot ought to appreciate the simple arithmetic at work here."

The Oriental lieutenant glared at Jon. Jon glanced at Fred. Inscrutable Fred stared back.

"I'm guessing you're Air Force Academy," Jon said.

"What's that got to do with it?"

"You believe the navy should just drive its ships and leave the flying to the air force. Here's a question for you. How does it help us fight the North Vietnamese if the US Air Force and US Navy fight each other? That arithmetic is even simpler than yours."

Scrutable Fred grinned. "That's it for the briefing, Navy. If it hadn't been a Saturday, it would've been a pleasure. Now I got a barbeque to get to. You'll have a sack lunch as you ride out to the site. Gourmet baloney sandwiches, I'm guessing. Oh, and smiling Leo is driving you out there."

"You air force folks know how to show a guy a good time."

When Stretch got back to Lemoore, RT directed him to a parking spot by the hangar. RT didn't want his maintenance personnel to lose both their Saturday morning and afternoon. He also had a message from LCDR Morrison. Zachery would depart for China Lake at seven Monday morning. Morrison wanted Jon to debrief him on the Nellis trip in the Weapons Training Facility at five on Monday.

Jon thanked RT for meeting him. "You have a retirement date set, RT?"

"No. The Bureau of Personnel promised us some folks next week. Others will be trickling in right up until you deploy. I told the CO I'd stick around for two more weeks. That's when the new ops O checks in. His name is LCDR Frost. He's at China Lake. Look him up when you're over there."

Jon thought he should say something, like *wish you weren't leaving*, but everything he considered saying sounded stupid.

RT smiled. "Go home to your family, Stretch."

As he drove toward base housing, the sun kissed the top of the Coastal Range, and warmth started leaching out of the air above the San Joaquin Valley. That chill was nothing compared to the reception he got from EJ upon entering his house.

Jennifer, however, shouted, "Daddy," and ran to hug him. Teresa invited EJ to do the same. "No!" he said. After a second invitation elicited the same response, Teresa swatted him on the seat of the pants and sent him to his room yowling.

At dinner, EJ wouldn't let Jon feed him. At bath

time, he wanted Teresa to bathe him, but she told Jon to do it, to more yowling. After EJ was dressed for bed, the boy discovered he couldn't yowl and suck his thumb at the same time.

For the Zachery family, Sunday went much better.

Chapter 12

Monday morning, Jon sat across from LCDR Morrison in what he thought of as the Little Round Top room, though it was probably against the law to even think that phrase. He'd just explained the cramped, claustrophobic confines of the Soviet-designed SAM radar and control vans.

"Seeing an actual site was worth a lot," Jon said. "I understand now that we're fighting the men in those vans, not the Fansong radar or the Guideline missile."

Morrison sat back. He frowned and sipped coffee from his mug. He used a knuckle to brush the bottom of his moustache. "Zachery, the weapon you will use is the Shrike missile. It homes on the Fansong radar. To fire the Shrike, you have to avoid getting shot down by the Guideline. It's the missile and the radar. How can you not see that?"

"I do that see that, sir, but the air force intel guys told me the North Viets have altered how they employ their radar and how they control the Guideline missile based on respect and fear of Shrike. When I visited the actual system, I could see why. I would not want to be jammed into the back of a cramped van with a Shrike missile coming out of the sky at me."

Morrison wiped a hand over his bald head, sipped coffee, and knuckle wiped his moustache. "Now listen, Lieutenant. We can argue points of philosophy later. Right now, just concentrate on learning as much as you can. Besides, Shrike employment tactics are above your pay grade. When you get to China Lake, you will get a brief on the TIAS system I told you about. You will get exposure to it and the CP-700 computer for firing the Shrike in a simulator."

"The CP-700 computer is a piece of crap," Jon said. "None of the airwing pilots trust it. It'll give you a couple of good bomb hits and toss the next one a mile away."

Morrison's pale cheeks sprouted color. "There were bugs in it. We got those worked out. It's essential for firing Shrikes. The launch envelope for Shrike is complex. Only a computer can weigh all the factors in an eyeblink and derive a solution to fire the Shrike with a high probability of killing the target."

"When RT and I had SAMs fired at us last cruise, we didn't use the computer."

Morrison leaned forward. "So are you saying you have the whole Shrike-firing envelope memorized? Why do you have to be such a difficult son of a bitch? Can't you see you've been invited into something new and innovative? Can't you see how important this could be to your career?"

"Career? That's what this is about? Careers? Why don't you call CDR Fuller and tell him you picked the wrong guy? Tell I'm too stupid to get it. Tell him to let me go back on leave."

"I can't do that. You've been briefed into the project. You've been to the Little Round Top site. Besides, I stuck up for you and convinced Fuller you knew something about SAMs. You owe me."

Zachery rested his forearms on the table. "You an East Coast puke too? You ever fire a Shrike? You ever had a SAM fired at you?" Zachery humphed. "I can see the answers on your face. Yes. No. No." He leaned farther forward. "You send me to learn something, and then I try to talk to you about what I've seen, and you won't listen. Your mind is made up, and it's closed. So, I'm going to fly to China Lake and see if I can learn new stuff you won't want to hear about."

Zachery stood. "And I don't owe you one dadburned thing. Sir."

Stretch pulled the throttle back and nosed over. Beneath him lay the eastern edge of the Sierra Nevada Mountains and few signs of human habitation. Farther east, the runways at China Lake stood out. The town of Ridgecrest snuggled up to the southern edge of the naval air station as if the life of the town depended on staying very close to the navy base. The rest of his view was barren brown. Looking out his canopy, Death Valley was at about ten thirty.

He had flown over the area at night and had been amazed at the number of lights dotting the floor of the desert. There was more life down there than appeared possible during the day, at least from four or five miles up.

Stretch checked in with the tower and was told to hold south and west of the airfield. "We have a missile test underway."

He rogered the tower and set up an orbit keeping him to the west of Ridgecrest. On his second trip around the circle, a ball of flame bloomed in the sky to the north and east of the airfield. *Sidewinder.* The test had been of a heat-seeking missile designed to home on the engine exhaust of an enemy fighter. The "Winder" had obviously hit the target, which had to be a drone airplane. The ball of flame plummeted toward the desert, trailing black smoke. Jon knew he was watching a drone fall from the sky. Still, it made him think of Skunk and AB. That's what their planes would have looked like after being struck by SAMs.

The tower cleared Stretch to land, which he did. Then he was guided to a hangar, where a sailor directed him to a parking spot. After he deplaned, the sailor told him Maintenance Control was in the hangar. He should check in there.

Jon stepped through the gapped-open hangar doors and saw a scene as alien as the moonscape outside. Rows of aircraft sat backed into spots against both sidewalls of the cavernous structure, but no two of the dozen planes were the same. An A-4, an A-7, an A-6, an F-4, and an F-8 represented models the US Navy currently operated in the fleet. Parked among the jets were propeller types from World War II and Korea. There was also a trainer airplane he had flown while earning his wings.

Jon was used to seeing twenty- and thirty-year-old petty officers in work uniforms or navy-issued coveralls working on planes. Here at China Lake, upward of fifty-year-old fogies comprised the aircraft maintenance force. No two of them wore the same sort of outfit.

One cluster swarmed over the F-4 on one side, while another half dozen worked on the Korean War vintage F-9 trainer. This one was a drone with a large black box of electronics filling the cockpit instead of an ejection seat.

He thought of the dying drone falling out of the sky fifteen minutes ago. Now maintenance techs were readying this drone to be shot down. The F-9 was parked opposite the F-4. Probably it would shoot down the drone tomorrow. Jon thought he was looking at some sort of airplane death row, and the F-9 was looking at its executioner.

"He'p ya?"

Jon had been standing just inside the doors, taking in the sights. A fellow, his own height and wearing paint-stained coveralls and a gray moustache, a mouth-covering soup stopper rather than a strainer, stood next to him. Big Moustache cocked his head to the side.

"Uh, where's Maintenance Control, please?"

BM snorted and nodded to Jon's right.

On the hangar wall, above the airplane tails, in monster-sized letters, it said:

Maintenance Control

"Oh."

Big Moustache slipped out through the crack in the doors. "Never seed no pilot needed no seeing-eye dog afore."

Inside Maintenance Control, Zachery checked in with a regular-navy-in-appearance chief petty officer behind the counter. Jon handed over the documents concerning aircraft side number 510.

"It'll take us a week to turn your plane into a TIAS bird, Lieutenant."

A sailor, clad in a dungaree uniform, entered.

"Seaman Neidlinger here will drive you to building 94, home of the TIAS lab."

"This way, sir." Neidlinger was short and looked like a Filipino. He sounded American. He was definitely a talker.

Before the sailor had the car out of the parking lot, Zachery learned the kid had come directly to China Lake from boot camp. China Lake was an armpit of the planet. "The guys say earth has a lot more than two armpits. You think that's so, Lieutenant? And would the navy send me to two pits in a row?"

Neidlinger asked questions, but he didn't wait for answers. The kid droned on like cicadas on hot summer nights back home. He drove them on a two-lane blacktop laid down amid scrubby brush destined to be tumbleweeds. Every so often, they passed flattened jackrabbit carcasses lying atop the road.

"Oh, yes, sir. Those are jackrabbits all right. You know what they say about rabbits, right? Well, we got a lot of them here at the Lake. Those flat ones we call sail rabbits. You can fling them like a Frisbee."

Evans pulled up to a multistory building and stopped talking.

"This it?" Jon said.

"Yes, sir."

"How do you know? There's no sign saying building 94."

"Sir, if you're stationed here, you know it's building 94. If you're a visitor, one of us Lakers would bring you here."

Zachery thought it was the first sensible thing he'd heard since the phone rang Friday morning.

Inside, a civilian security guard checked his ID, pointed to an elevator, and said, "Fourth floor, room 401."

Outside 401, he pressed the buzzer button and then was admitted by a white-coated lab tech and led to two men standing beside the cockpit of flight simulator.

One of them, a six-foot navy LCDR in khakis, wore a "Frost" nametag. "Will Morrison told me about you." Frost didn't offer a handshake.

The other was Zachery's height. "Ross Hill," he said, and stuck out his hand. After a vigorous pump up and down, he said, "This way." He spun on his heel and led them into a conference room with a projector and viewgraphs.

Zachery took a seat, and Hill turned on the projector. The slide said, "This briefing is *secret*." Hill stared through thick-lensed, black-framed glasses at Jon. Then, apparently satisfied that his audience received the message, he said, "Here's the agenda for today, Lieutenant. I'll give you a brief on the TIAS system. LCDR Frost will talk to you about flying our simulator to familiarize you with the system so it's not strange to you when you fly it on the range tomorrow. This afternoon, you'll be driven to our own mock-up SAM site so you can watch an A-7 fly simulated SAM encounters and simulate launching Shrikes. You'll see the test set up and how we collect data."

Hill then launched into a description of the TIAS system, the components, the placement of antennas, and the switches and display in the cockpit. Mr. Hill spoke with enthusiasm and delivered his lines at a measured pace. Jon wondered if he'd been a teacher, a professor maybe in some former life.

Hill said, "I included some details I thought you'd appreciate, Lieutenant, being an electrical engineer. I generally leave those things out when I'm speaking to phys. ed. majors." Here he aimed his glasses at LCDR Frost and sat down.

Frost walked forward to stand by the projector.

Frost said, "Our Ross Hill has beans to spill but only about electronic boxes, which he, in his lab, *concoxes*."

Zachery was the only other person in the room. Odds were Frost majored in phys. ed. and his literary masterpiece was aimed only at Mr. Hill.

"Ross is the TIAS project engineer," Frost said. "I'm the project test pilot."

Zachery thought, *And a self-inflating balloon of pomposity.*

The LCDR showed viewgraphs of the Shrike envelope, different altitudes and ranges at which a Shrike could be fired with a decent probability of killing a SAM site. Next he showed a picture of the TIAS display, which was the radar scope in the A-4 cockpit.

"Ross deserves some credit," Frost said. "His design, when coupled with the CP-700 computer, results in more accurate Shrike firings, and here's the real kicker. You can get a Shrike off in half the time it would take you with the standard Shrike firing method."

"The key factor," Hill interjected, "is the TIAS display is linear. SAM targets show up as a dot on the radar scope. If, for instance, the dot is ten degrees left and ten degrees down, that is precisely where the target is."

"Right." Frost looked annoyed. "With a TIAS bird, you turn and dive smartly to center the dot on the radar scope, mash the bomb button, which locks the firing solution into the CP-700. Then you honk back on the control stick, pulling the nose up until the computer fires the Shrike. Takes half the time as the old way."

"The linear display is the key," Mr. Hill said. "In the old system, target sensors in the nose of the Shrike missile drive horizontal and vertical needles on your cockpit display, and since the old display wasn't linear, the pilot had to spend extra seconds getting the needles precisely centered."

"Okay, Ross, let's stick Lieutenant Sea Dog in the simulator and see what he's got."

"Sea Dog?"

"Right, Lieutenant. You just came back from a deployment. We expect you to show us shore duty pukes how it's done in the fleet."

Zachery thought, *Setup!* Odds were Frost intended to humiliate him.

"You coming, Sea Dog? Time's a wasting."

In the simulator room, there were five separate full cockpit sims, including the current fleet fighter and attack models. Frost was waiting for him by the A-4.

Stretch climbed in.

"Okay, Sea Dog," Frost said. "Your setup will be twenty thousand feet, speed Mach .7, a target will appear on the TIAS scope. Turn and dive to center the dot and fire a Shrike just like I briefed."

The simulator hummed. Cockpit lights came on. The needle on the airspeed indicator sprang to Mach .7. The altimeter registered twenty thousand feet. A pinpoint dot appeared on the scope ten degrees right and eight degrees down. The threat warning system indicated he'd been locked up by a Fansong radar.

Stretch banked and dove, overshooting badly in turn and dive angle.

"Freeze it," Frost said.

"Well, Sea Dog, you overstressed the airplane, and it was going to take you forever to center the dot and shoot a Shrike."

"Give me the initial setup again. I need to play with the sim a few minutes to get the feel of it."

"We don't have time for this, Sea Dog."

"Really, Commander? How long did it take you to learn how to fly this thing the first time?"

"It took him an hour," Mr. Hill said.

"Ross, get the hell out of here. This is my show."

"Negative," Stretch said. "If only one of you is staying, I want it to be Mr. Hill. I trust him. All you wanted was to show me up. You've done that. Your work is done here, Commander."

Frost cursed, stomped to a phone, and dialed a number. When the other party picked up, he said, with his voice raised, "Zachery's just like you said, Will. Arrogant, won't listen. It's a waste of time trying to work with him."

Morrison wanted to speak with Zachery. Zachery listened to a ninety-second tirade, and when it ran out of steam, he said, "Okay, Commander, now listen to me for thirty seconds. LCDR Frost hasn't taught me a dadburned thing. All he wanted to do was humiliate me." He explained about the simulator and not having any time to develop a feel for the machine.

"Put Frost back on." Morrison sounded miffed.

Teresa had just gotten the children to bed, and she'd returned to the kitchen to wash the dishes. The phone rang. She answered.

"Mrs. Zachery. Will you accept a collect call from Mars?"

"Silly. Mars. I'm so sure you can go there in an A-4. I just put the children to bed. I'm going to—"

Little feet pattered into the kitchen. She let EJ talk first. Amazing. The boy was pleased and happy to talk to his father over the phone. He wasn't pleased and happy to deal with his father in person. Teresa put her hand over her heart.

Chapter 13

On Monday afternoon, US Navy LT Buzz Lemoine drove Zachery to a mock-up SAM site. The mock-up vans were nothing like what Jon saw at Little Round Top. These vans were spacious, affording plenty of room for spectators. The displays and monitors appeared to be new. The Soviet electronics, to Jon, looked to be as old as his pop. But according to Buzz, the site replicated the signals of the Fansong radar and missile guidance system well enough so the threat warning systems in the airplanes couldn't tell the mock-up from the real thing.

While they'd been at the mock-up, they observed an A-7 make several runs against the site and simulate firing Shrikes. The preliminary data collected by the range indicated all the shots had resulted in a SAM site kill.

After the A-7 completed its tests, Buzz drove them back to the base.

Jon asked him about what they'd just seen. "The pilot took a minute to get a Shrike off after the site brought up the Fansong radar signal. In the real world, the guy would have been dead before he fired."

"Yeah. But at this stage, we're only testing how well the A-7 computer works. That airplane has the newest version of the software. We want to know if it delivers the weapons accurately. And it does. The flight we saw completes all the test points from a twenty-thousand-feet firing altitude. I fly tomorrow morning and fire from fifteen thousand."

"And you don't care how long it takes to fire, and you don't care that all your runs are right at the site?"

"Like I said, we are trying to prove the software does its business correctly. The guys at VX-5 in another hangar on base, they worry about using it in combat."

As he drove, Buzz seemed as starved for a sympathetic or captive ear, as Seaman Neidlinger had been. Besides Al Frost and Buzz, the other pilots in the Projects Department were civilian and retired military. "Old farts, you know? And Frost, he isn't a fart, he's a regular shit."

Buzz found out that Al Frost had reported to the USS *Oriskany* in the Tonkin Gulf in the middle of a deployment as a replacement pilot. "He flew one hop from the O Boat, which was a tanker. The next day, that big-assed fire broke out, and O Boat and Frost returned to the States. So he never poked his nose over North Vietnam. Al went to an East Coast squadron and did a deployment to the Mediterranean. He met Will Morrison in that squadron, and they became buds. After that Med cruise, Morrison went to the Pentagon, and Frost went to a training squadron. After being an instructor pilot, Frost was ordered here to the Lake. And guess what?" Like with Neidlinger, his questions did not ask for answers. "When Al checked in here, he told everybody his call sign was Combat Al."

Buzz talked to guys on the East Coast who knew Frost's history. Buzz blew the whistle on Frost at the Officers' Club one Friday afternoon. After that, the retired-old-fart pilots gave Al the cold shoulder. Now they all called him Combat Al. "But before I blew the whistle, they never called him that."

"So, he's your boss, right?" Zachery said.

"Not for long. Real soon, he's going to be your boss. Congratulations, by the way."

Tuesday morning, Stretch and 510 took off and headed for his firing run starting point at twenty thousand feet and twenty miles from the simulated SAM. After turning south, he checked his altitude and airspeed and called, "Five ten inbound."

Seconds ticked away. He thought, *Okay. Anytime now.*

His threat warning system buzzed: a SAM radar was locked on. Even though he knew it was coming, it spurted beat-fast juice into his heart.

What in Sam Hill?

His TIAS scope showed a blop as big as his thumb. In the sim yesterday, a pinpoint dot designated the location of the target. Stretch pushed his nose down and tried to center the blop, but it wasn't round, more like a smeared thumbprint. He finagled his dive angle and decided it was as centered as he could make it. Mashing the bomb button, he pulled the stick back until the system signaled he'd fired his Shrike.

As he turned and climbed toward the same starting point, he called, "Range Control, this is 510. When I turn in on the next run, bring up the target when I tell you, and leave it on until I say take it down. Over."

"Roger, 510. If that's what you want."

"Negative, 510." *Frost.* "We're burning project funds here. We need you to get data points. You learn along the way."

"If you'd have told me about the difference in the sim display to the one in the airplane, I would have learned something on the first run. As it is, the sim session yesterday was a waste of time. Now, Range Control, do what I said. Keep your data recording going through the entire run."

"Zachery," Frost again, "fly the mission the way I planned it."

"Range Control, 510 is turning in for the second run. Do it the way I said, or I'm returning to the field, and Commander Frost can explain why we wasted a whole range period because of his fouled-up plan."

Frost didn't appear on the radio again, and Stretch flew five runs against the SAM, with the range doing it his way. After he landed and did his postflight inspection, he said, "Thanks, 510." He turned and spotted Big Moustache standing in front of the hangar. Zachery acted as if he didn't see him and walked into Maintenance Control.

The chief pointed to a phone handset lying on the

counter and said, "A pissed-off Lieutenant Commander Morrison wants to talk to you right away."

Zachery picked up the phone and said, "I'll call you later, Commander." Then he hung up and dialed another number. Wazinski answered the phone and put Jon through to the CO.

Fuller launched into a tirade. Zachery waited until he paused for a breath and said, "Commander Fuller, just shut up and listen for a minute."

Fuller tried to reassert control, but Zachery blurted on. "If you want TIAS to amount to something, we need somebody over here running things other than LCDR Frost. He's a peanut-brained juvenile, and he's not doing TIAS any good. Now, I need a couple of hours to talk to the project engineer in the TIAS lab, and then I'll call you back. Will you give me that, sir?"

Fuller gave Zachery career death threats first, for talking to his skipper the way he had, but, second, he gave him one hundred twenty minutes. "And the clock's ticking."

Zachery checked his watch, looked up, and told the chief that 510 was a good bird. He had no gripes to report.

"Mr. Z," the chief said, "listen, I'm sorry for what Roscoe Mudd painted on your plane. He told me he was going to paint your name on it, not See Dog. I'll have him paint that out."

"Nah, Chief. Don't do that, please. Leave it the way it is. And I need a ride up to building 94 like chop-chop. I only have one hundred nineteen minutes left."

In the simulator room, Ross Hill explained that he had designed a display for TIAS. It cost a little more than modifying the A-4 radar scope, but CDR Fuller and LCDR Morrison had decided it would be easier to sell the program in the Pentagon if they touted reusing existing hardware rather than buying all brand-new equipment. His design was what was installed in the sim.

"Will the sim work with the A-4 radar display?" Jon said.

"Sure. It takes ten minutes to swap out," Mr. Hill said.

"Can you do it now, please? I have to call my CO back in eighty-seven minutes."

With the blop providing target location information, Zachery's kill rate in the sim fell to 50 percent.

"There's one difference in how the display works in the sim and the aircraft. In the plane, once you get that blop close to centered, the middle of the blop kind of sparkles, in effect reducing the size of the blop. I didn't see it on my first run, but I saw it on the second."

"Huh," Mr. Hill said. "The range guys gave me preliminary results from your runs. You got kills on every shot except the first one. The range guys don't like to give project pilots prelims. If the final results aren't as good as the initial cut, the pilots get torqued off at the range guys."

Zachery climbed out of the sim cockpit with ten minutes to spare. "Mr. Hill, check me on this please. LCDR Frost flew the TIAS sim with your specially designed scope because he got over 90 percent kills that way. Out on the range, when he had to deal with the blop, he got less than 50 percent kills, but from when they brought the target up, all his shots were in thirty seconds or less. That was a key selling point of the program. According to what you said, he blamed his low kill rate on either the range or a problem with the TIAS system, but you don't think there's anything wrong with it. Or with the CP-700 when it shoots Shrikes. You can't talk to how it does dropping bombs, but Frost, if he'd practiced in the sim with the regular A-4 radar scope, he'd have gained familiarity with the blop and probably improved his kills."

Hill agreed Zachery had it right.

"Why don't you call LCDR Morrison, Mr. Hill, while I call my CO."

"Frost wouldn't let me talk to him."

"Frost isn't here. I think you should call Morrison."

Zachery dialed with five minutes to spare. His CO greeted him with, "Well!"

Jon rolled out his observations, the things he learned from Ross Hill, and his conclusions. "LCDR Frost is only interested in making himself look good, not in proving if TIAS works or not. TIAS is a good system. It'll take a couple of days for the results from my runs to be finalized, but Mr. Hill thinks we are finally on the right track. I am going flying again this afternoon, and Mr. Hill and I worked up improvements to LCDR Frost's test plan. I can call you again after that if you want me to."

"Call." Fuller hung up.

On his afternoon flight, Stretch and 510 achieved Shrike firing times of less than ten seconds. After he landed, he was driven to building 94, where he compared notes with Mr. Hill.

"Preliminary results say you got 100 percent kills with your shots." Hill delivered this news deadpan. "If the final results show even 85 percent, that will complete our proof of concept testing. Exciting stuff, eh, Lieutenant?"

"Oh, yeah, Mr. Hill. And I'd like to say you built one heck of a fine system."

Hill straightened up as if he'd just received some data he couldn't process.

Jon asked Hill to get on the phone with him to his CO.

Fuller's brusque antagonism came through the speaker phone loud and clear. Mr. Hill looked at Zachery and rolled his eyes, which as seen through those thick-lens glasses surprised a smile onto Zachery's face.

Hill delivered his assessment on the preliminary results of Lieutenant Zachery's test flights. They waited for Fuller to say something, but he didn't.

"Commander, TIAS is a great system. But I need to be able to look at some offset encounters, not having each flight setup to fly right at the target SAM."

Fuller said, "Call Morrison and tell him what you need." The dial tone droned.

The airspace over China Lake was segregated into a

number of different ranges so that simultaneous tests could be conducted. Mr. Hill called the man who scheduled access to the various ranges. When he hung up, he said, "At the times we need it, the entire airspace is available to us."

They called Morrison. Morrison was pleased that it looked like they had completed the proof of concept testing with money remaining in the budget for training Warhorse pilots. "If the additional range is available, go ahead and schedule it."

The next day, Stretch and 510 began their test runs from the western boundary of the China Lake range. When the target SAM came up, Stretch turned, centered the blop, and simulated firing a Shrike with the CP-700 computer and achieved a 90 percent kill rate with a twenty-second-or-less time to fire. In the afternoon, he repeated the routes he'd flown, except he fired simulated Shrikes in manual mode without using the computer, which meant he did not have to turn and dive to center the blop, then pull the nose up to launch. Rather, in manual, he could turn and pull his nose up directly to the required launch angle. His launch times in manual were back down in the ten-second time to shoot area.

After he landed, he called Morrison again and told him he'd gotten excellent results even with the offset encounter angles, and the CP-700 was a great asset for launching Shrike. He didn't tell him of the success he had in manual mode firings.

"Also, Commander," Zachery said, "I am done with what you sent me over here to see. With the revised test plan Mr. Hill and I devised, I actually learned a lot more than I would have. And I've got a great idea as to how to train more Warhorse pilots in how to use TIAS. We'll use the simulator here at China Lake for an introduction to the system. Then, each time the Lake is finished installing the TIAS system in a new airplane, we'll have a different pilot come over and fly the TIAS checkout flight. TIAS is a heck of a system, sir. The Warhorses will be the best SAM fighters anywhere."

Zachery winced. That level of brownnosing a guy like Morrison left an aftertaste.

"One more thing, Sir. I've talked to Mr. Hill about using a Lieutenant Lemoine here at the Lake about helping train our pilots. He'll do a much better job than LCDR Frost. Uh, in my opinion, sir. By the way, I haven't seen Mr. Frost around today. Do you know where he is?"

"Sure. He's here at Lemoore. Matter of fact, I think he's talking to your skipper as we speak."

That morning, Teresa discovered puffy cheeks. After she fed the children lunch, she discovered puffy ankles. Now, at three o'clock in the afternoon, she lay atop their bed, pillows propping her up as she read and waited for Jon to call. Last night, he'd said he might have to stay an extra night.

She already knew he should have just skipped the *might*.

The phone rang. She answered after one ring, said hello, listened, and said, "You have to do what you think is right."

The children ran in for their turns with Daddy. When they were finished, Jennifer handed over the phone, and they ran back to the living room.

"You have to do what you think is right, but I sure wish it had been right for you to come home."

"If only doing the right thing didn't cost me nothing and you and the kids everything."

"Jon Zachery, you know how much I love you."

"No, I don't, but I'm very interested in being convinced tomorrow night."

Jon had called the base at Lemoore via the government phone system. The Lemoore base operator patched Teresa to Jon's call. It was nice to not have the length of a roll of quarters dictate the duration of their conversation. Technically, it was against the rules to make personal calls over the government system, but Teresa knew RT called Helen that way at times. Never at periods of high government call volume. Compared to what the US Navy had done to the Zachery family the past week, one measly phone call wasn't a big deal at all.

It might not be the absolute right thing to do, but it was okay.

When she laid the handset gently on the telephone cradle, she wondered, as she did often, *Was this the very best phone call with my high school sweetheart ever?*

Jon donned running gear, and, following instructions from the sailor behind the desk at the BOQ, he ran a two-mile route out and back over a road graded through future tumbleweeds. Creosote bushes, according to the petty officer, and as the PO cautioned, he stayed in the middle of the graded surface, well clear of the edges where snakes might be resting in the shade. He pushed himself to sweat out two lieutenant commanders and one commander. He didn't want them inside himself where Teresa abided.

That night, as he was about drift away, his eyes popped open. It was clear he'd divested himself of the two LCDRs but not the full commander.

Crap, he thought.

Thy will be done, he thought.

Then he thought, *What? That's what You were waiting for?*

Chapter 14

As Stretch taxied 510 between rows of parked A-4s, he spotted a woman and two children next to a LCDR in khakis, standing next to the Warhorse section of the hangar. *Teresa?* The navy guy was bald, obviously not wearing a hat on the flight line. *Morrison?*

No. This guy did not have a moustache, and he wasn't much taller than—Teresa? Definitely Teresa.

He parked, shut down, and climbed out. RT was waiting for him by the nose of the aircraft.

"Stretch," RT said. "Fuller's taking you to admiral's mast in the morning."

"Mast? What for?"

"LCDR Al Frost told me you blew through the entire TIAS budget while you were at China Lake."

"Nobody told me anything about a budget. Everything I did I had permission to do. In the simulator, Mr. Hill, the project engineer, okayed what we did. On the range, I told the CO what I needed. He said ask Morrison. I did, and Morrison gave us the okay to modify the test plan."

"Frost said he was the guy who managed the China

Lake test budget. Morrison never had anything to do with that. When you bumped Frost out of the project, you also removed project funds management. Fuller blames you for that and for blowing the budget."

Stretch turned toward the hangar. The LCDR had Jennifer by the hand. Teresa had EJ. Jennifer looked like she was towing her escort. It appeared Teresa was half dragging EJ.

Jennifer pulled her hand free and ran to Jon, who dropped to one knee. Jennifer threw her arms around his neck, squeezed, and said, "Urk!" She released him, and Jon said, "Ah."

Jennifer stepped back, smiling big.

Teresa pushed EJ, but he moved only a step. "EJ!"

EJ threw his arms around his father and squeezed, surprising an "Urk" out of Jon.

"No!" EJ said. "I urk."

Jon got it right the second time; then he stood, looked into Teresa's eyes long enough to register the depth of the worry and the hurt in them, and hugged her gently.

When they parted, RT introduced LCDR Dave Davison, the new XO. "He checked in yesterday."

The first thing Jon noticed was the XO had puffy cheeks and a bit of a belly.

They shook and were happy to meet each other.

The XO glanced at Teresa and back at Jon. "You know what's happening tomorrow, right?"

"The CO wants to take me to admiral's mast."

"Right, so be out here at 0700 in whites."

"I'll be here in whites, XO, but I refuse mast. I demand a court-martial."

"Court-martial!" Teresa's soprano and the XO's baritone delivered the two-word aria perfectly harmonized.

"You can't do that," the XO said.

"Call the base lawyer, XO. He'll tell you I can."

"Talk some sense into him, RT."

"I think Lieutenant Zachery knows exactly what he's doing, XO. Besides, once he's got his mind made up, he's a Missouri mule."

The XO shook his head. "Welcome to the Warhorses, LCDR Davison. We are one big happy family, and we hope you like it here, LCDR Davison. See, we even have a coffee cup for you with your name on it, LCDR Davison. What a bag of—" He glanced down at the children.

"My suggestion, XO," RT said, "is to call the lawyer. See what he says. Then call the skipper. Then one of you will have to call the admiral's chief of staff."

"Zachery, please take your family home. I have a fierce need to cuss like a sailor."

"Yes, sir. Sorry, XO."

"Go!" The XO's eyes looked like they might pop out of their sockets.

Jon picked up EJ. RT picked up Jennifer. Teresa took Jon's arm, and they entered the hangar.

RT said, "Teresa, I need to get a few facts from Jon, okay?"

RT explained that his understanding of the situation was that when Jon asked to use the extra air space at

China Lake, nobody told him the range charged a lot more to lock down the whole area.

"Nobody told me. I wasn't smart enough to ask."

RT then wanted to know how the permission to use the extra air space had come about. Who did Jon talk to and when, and were there any witnesses?

"The chief petty officer in maintenance control at China Lake heard me talk to Commander Fuller. On Tuesday. Fuller told me to call Morrison. I called Morrison with the TIAS project engineer, Mr. Ross Hill, on the phone with me, and Morrison told us we could sign up for the extra air space."

"I'm going to do some digging. I'll stop by your house on my way home. Write up a timeline on what happened during your stay at the Lake, would you?"

They were in bed, lying on their backs. They held hands.

Teresa stared at the ceiling, a sky of nightlight lit gloom. "I'm afraid, Jon."

"Me too, sweetheart. I'm afraid of what this is doing to you and Little Pootzer. The navy makes it tough on me, but it imposes agony on you."

"Jon Zachery?"

"Yes, dear."

"Would it be too much of an imposition to ask you to put your arms around me?"

They jiggled themselves onto their sides.

Jon said, "Mmm, why didn't I think of this?"

"Aren't you afraid of what might happen to you? And why did you demand a court-martial?"

He kissed her, and it was close-your-eyes nice.

"As to what will happen to me, I did my best to see the aspects of what was in front of me. I did my best to discern the right thing to do, and I did my best to do just that. It was the only thing I could do."

Just when she decided he wasn't going to answer the other question, he did.

"Maybe I should just trust the admiral to sort all this crap out. But I don't know him. I've seen him at a few formal functions. But I don't know him well enough to say whether he's a good guy or cut from the same bolt of cloth as Fuller. As an ensign, I served on court-martials. When I did, we worked hard to get all the facts in hand before making a judgment. All I can say is court-martial seemed like the right thing to do."

Teresa mulled the pluses she could see with a court-martial. It didn't take long. The minuses ran out of counting fingers on her two hands.

"Teresa."

Jon's tone of voice made her suspicious. She raised up on an elbow.

"This afternoon at the hangar, did you notice that the XO's cheeks were puffy?"

"Jon Zachery!" She slapped him on the chest. It made a very loud *smack!* "I'm sorry. Are you all right?"

"No. My heart stopped. Only mouth-to-mouth resuscitation can save me."

The next morning, at ten before seven, Zachery entered admin. The CO's door was closed, the XO's open. Puffy Cheeks waved him in, but Zachery stopped in the doorway. The XO whispered, "Our new CAG is in there with the skipper. His call sign is Bear. He looks like one and growls like one too."

The CO's door opened. The guy looked like a bear all right, and Tiny might not be the biggest pilot in the airwing anymore.

CAG said to Wazinski, "YN1, Zachery's supposed to be here shortly. See if he's in the ready room, please. If he's not there, leave word that he's to hustle right here. And it'd be best if he isn't late."

"He's right there, CAG."

The big, low-browed head turned. The dark eyes swept down and up. "You're kind of a little shit to be stirring up such a big grunch of trouble." After a moment, he said, "Nothing to say for yourself?"

"Sir, you didn't ask me a question. You made a statement with two facts in it. I can't argue with either one. I am not a big guy, and I had a hand in stirring up trouble."

"And a wiseass."

"I don't think I am, sir. You pressed me to say something, so I did."

CAG's gaze swept over Zachery again. The left corner of his lips tweaked up a smidge.

"Okay, Lieutenant, come in the skipper's office and explain how it is you know more about fighting SAMs than he does."

Commander Fuller sat behind his desk. CAG took another chair.

If looks could kill. Fuller glared at Jon, and this time, Jon glared back.

"Knock that shit off, Fuller. I did not get in an airplane at oh dark thirty so I could come up here and watch you engage in a stare down contest with a JO. Zachery. Fighting SAMs. Let's hear it."

Surface-to-air missiles were part of an overall air defense network. The North Vietnamese SAM operators used the network to get targeting information on American planes in a way that minimized their exposure to the threat of a Shrike missile.

Zachery said, "The SAM flies for forty seconds to intercept, and as soon as they kill or miss a target, they take their radar off the air. A Shrike will fly a minute twenty seconds to reach the SAM site. Even if you fire a Shrike, it will have nothing to guide on for the last half of its flight."

"The same old crap he's been spouting all along, CAG. The issue is the Shrike firing envelope is complex. You need a computer to fire the Shrike with a chance to kill the SAM."

"But you won't kill the SAM because the SAM radar is offline."

"So what are you saying, Zachery? That it's hopeless to even try to fight the SAMS?"

"Absolutely not, sir. The North Viets respect Shrike. That's why they operate their system the way they do. If we can get all the bombers in a strike to their target, isn't that worth firing a couple of Shrikes even if we don't kill the SAM site?"

"What do you think, Fuller?" CAG said.

"CAG, he's had one SAM fired at him, and he's the world's expert?"

"Nothing of what he said means anything to you?" CAG said.

"Uh, CAG, uh—"

"That wasn't a trick question, Commander."

CAG asked Zachery to wait in the ready room.

As he got up to leave, there was a knock on the door.

"Not now," Fuller shouted.

"Open the damned door." CAG was louder.

Wazinski stood in the doorway. "CAG, the admiral would like to see you in his office. He's sending his car for you."

"We'll continue this later," CAG said to Fuller; then he followed Zachery out into the passageway.

"Zachery, at this point, I'm sure you know a hell of lot more than your CO and those two LCDRs he's bumbling around with, but you do know that when you bumped that Frost guy off the program, he was the only one with knowledge of test range scheduling and budget."

"I know it now, sir. I saw a way to fix what was screwed up, and I went after it. I tried to talk to the CO and the LCDRs, but not one of them would listen. Plus, well, my peanut-sized brain was too full of other stuff to see there might be consequences to getting Frost out of the way. Yes, sir, I screwed that up, big-time."

They stopped outside the ready room.

"Listen, Lieutenant, when I'm with the admiral, I bet I can talk him into dismissing the charges. We'll put a letter of caution in your record, with a probationary period of one year. If you don't screw up anything else in the year, the letter will be pulled. It won't affect your career."

Zachery thought about his family, about Little Pootzer. This was a way out, so he wouldn't bring any more grief down on their heads. *Temporarily.* If Fuller remained as Warhorse CO, he'd get revenge. He'd find a way. Grief would still find its way to Teresa and the children.

"CAG. Thanks, but if you're asking me to agree to this deal, I can't."

"Oh, for Christ's sake! Why the hell not?"

"A part of the deal is CDR Fuller remains in place, right? Well, if he stays, I am 99 percent certain the anti-SAM mission in *your* airwing, sir, will not be worth much."

"I'm not sure how you going to court-martial fixes that."

"Me either, sir. But it's all I got to throw into the game."

"You know, Zachery, yesterday at my change of command, when I became the airwing by-God commander, I was sure pleased with myself."

"May I say something, sir?"

"I bet this is going to be good."

"It's my observation, sir, that the state of being pleased with oneself is an ephemeron."

"Jesus, God in heaven, save naval aviation from pilots who read books."

CAG walked away. When he got to the stairwell down to ground level, he turned and looked back at Zachery and laughed.

They were on their backs in bed and holding hands.

Teresa said, "From court-martial to Navy Commendation Medal, you've had quite a day, Jon Zachery. And I'm so glad the admiral invited the children and me to attend the presentation of your medal."

She squeezed his hand. "'For superior performance of duty.' Of course, there was the yada-yada that came before it. 'Lieutenant Zachery discovered flaws in the methodology used to train TIAS test pilots. He devised corrected training procedures as well as improved procedures for the conduct of the actual test flights. His efforts led directly to proving the TIAS concept, requiring only half of the planned test flights.' And then more yada. I am so proud of you, Jon Zachery, and of each and every one of those yada yadas."

"You are the one who really deserves a medal. I just did some pilot stuff, but do you know what you do with your Teresa mysteries of the rosary?"

"Pray?"

"Yes, that, but pray meaningfully. Every once in a while, I find myself praying prayers while my mind wanders. It's like I've wrung the meaning clean out of the words by saying them over and over. But your mysteries, they are personal, packed with in-the-moment meaning. You save my soul, Teresa Velmer Zachery."

"Are you saying, without my mysteries, your prayers are like chewing gum that loses its flavor on the bedpost overnight?"

"Mrs. Teresa Velmer Zachery, this is a solemn and serious moment. Because I hereby award you a commendation medal for superior performance in the field of prayer-make-upping. Now, you wait right there. I'm going to get a medal, which is right over there on the dresser, and pin it to your nightgown."

When he came back, she said, "Please don't stick yourself with that thing. Or me. You can pin it to my nightgown later."

Later, Jon slept peacefully beside her. She liked being beside him as he rested, with all his cares and woes lifted. *For the moment*. But they had the moment, and it was one to savor. Especially considering how the day began.

She felt herself sinking but clawed her way back to wakefulness. The medal on her nightgown could hurt Jon or herself; but then she decided to leave it there.

PART 3

WELL, YES, THIS IS NORMAL, BUT IT'S NOT NORMAL-NORMAL

Chapter 15

After the Zacherys visited Dizzy Land and Monterey, they returned to Lemoore on the same day Mike and Amy Allison returned from leave in Iowa. The next morning, Mike drove himself and Jon to the squadron.

"Talked to RT last night," Mike said. "He told me you were going to admiral's mast, wanted a court-martial instead, and then wound up with a Navy Commendation Medal. How in Sam Hill did all that happen in a week?"

"I planned to tell you. Just didn't know where to start."

"How about the beginning?"

Which was EJ's broken heart. Jon related the abrupt cancelation of his leave, his experience with US Air Force intel guys—leaving out the compartmented stuff—and meeting Ross Hill, the TIAS project engineer. He talked about CDR Fuller and his two stooges, Morrison and Frost. At first, Morrison saw value in Jon's understanding of the anti-SAM mission, but Fuller didn't want knowledgeable subordinates. He wanted, "Aye, aye, sir. Anything you say, sir," from them.

When Jon asked for access to more China Lake

airspace to conduct test runs, they saw an opportunity to get rid of him. Frost knew the extra space cost more than the narrow slice they'd been using. Utilizing that extra space would run the project out of funds. They'd blame it on Zachery. When Jon orchestrated the removal of Frost from the project, he removed the only person with knowledge of the project budget and range costs. Fuller was certain that once they got rid of Zachery, he could weasel more money out of the Pentagon to keep the project going. But, to keep all aspects of the program under control, Zachery had to go. He was a loose cannon. He would not keep his mouth shut. He had to go.

"And this scheme concocted by the three stooges," Mike said, "it all fell apart when RT found this logbook outside a secure briefing room in the Weapons Training Facility. The logbook and the testimony of a marine lance corporal proved Frost and Morrison were in the same small office when Morrison gave the TIAS Project permission to obtain the extra airspace. So the whole basis for their claim against Lieutenant Zachery fell apart. That right?"

"Pretty much. Once RT found that, he and the admiral's lawyer questioned the stooges one at a time. Fuller and Morrison didn't break, but Frost crumbled."

Mike shook his head. "But how would they expect a fleet lieutenant to understand about budget and range costs when neither Morrison nor Fuller knew that stuff?"

"This is just what I think. The admiral and CAG didn't want me to go to mast. They tried to find a way to derail it but still throw CDR Fuller something. Like a probationary letter of reprimand for me. It would go in my record, but if I kept my nose clean for a year, the letter would be removed. The heavies thought they could give Fuller a stern talking to about how to run his squadron and how the anti-SAM mission should be run and that everything would blow over. They were impressed with Fuller's idea about individual squadrons specializing in the newest weapons, so they wanted a way to keep him in the game and not have me pay too big a price. Any rate, once RT found that log, and it was obvious Fuller and his LCDRs had run the project out of funds and blamed it on me, that changed everything."

Mike pulled his car into a spot by the hangar. "What happened to the stooges?"

"According to RT, they resigned from the navy about the time my mast had been scheduled, and the three of them, when they left the admiral's office, wore no rank insignia, no wings, and no ribbons on their uniforms."

They got out of the car, and as they walked toward the hangar, Mike said, "So, it's over?"

"Right, and now everything's back to normal."

"Stretch?"

"Yes, Alice."

"You do know that *navy* normal is not normal-normal, don't you?"

Stretch humphed.

That morning, the new skipper checked aboard: CDR Fant, Leroy, call sign Little Lord. He'd been slated to begin training to fly the A-7, but the Bureau of Personnel diverted him to fill an immediate need in VA-92 Warhorses.

The CO, the XO, RT, Stretch, and Alice comprised the squadron officer roster out of an allowance of nineteen. Half the enlisted billets were still filled. A sparse-looking crew assembled on the hangar deck to hear the CO read his orders and assume command, which he did, and then he said, "Boys and girls, there's a lot of work to do and not many of us to do it. Hop to it."

After the formation was dismissed, the CO, XO, and RT got on telephones to detailers in the Bureau of Personnel. Jon took Mike over to the Weapons Training Facility. Little Round Top room was now just a room cleared for material up to *ho-hum* secret. Inside, he showed the TIAS slides that had been signed out to Morrison. Afterward, Jon signed for the slides and brought them back for inclusion on the VA-92 list of classified documents.

At the squadron, Mike Allison stopped in the ready room, and Zachery continued on to admin, where he gave the slides to Twombly.

"Don't lock those up," the skipper hollered from his office. "Lieutenant Zachery, step in please.

"Change of plan for tomorrow, Stretch. CAG is flying into China Lake in the morning. He wants to observe Alice in the TIAS sim. So, instead of Alice going to the Lake by himself, you and I are going as well. Alice will still do the program the way you planned, but CAG says he needs to understand the anti-SAM mission and how we'll fly it. Right now, CAG understands what

TIAS is and how we'll use it better than I do. So I'd like you to run me through what you showed Alice."

The skipper had stopped talking. "Uh, yes, sir. You mean right away?"

"Any reason to wait?"

"Right now, the slides are 90 percent TIAS system description and only a bit on how we'll use it. I was planning on adding slides to cover that."

"How long were you figuring it would take you?"

"Well, Skipper, I figured I'd work over the weekend and have them done by Monday."

"It's eleven. Brief the XO and me in two hours."

"Any new slides I get done will be rough."

The skipper waved his hand. "RT says you are the best guy to lead the anti-SAM mission. He says Mike Allison will make a good deputy, but two senior lieutenants are checking in next week. Both logged combat hops over the north, both earned great reps. Before deciding on a deputy, we should check these guys out."

After a pause, the skipper said, "There's an awful lot of thinking going on for something that seems simple, straightforward, and dripping with common sense."

"Skipper, I've flown with Alice as his lead and as his wingman. I trust him, and after I briefed him this morning, he gets the mission."

"What about me, Stretch? Do you trust me?"

I trusted Tuesday and Botch once. Fuller and his LCDRs, I never trusted them.

"I know you're not CDR Fuller, Skipper."

"But you've trusted people before, and they betrayed that trust?"

"Yes, sir."

The skipper grinned. "Good. I was afraid you were just going to say you trusted me—you know, blow sunshine up my ass, tell me what I obviously wanted to hear."

"Uh." There didn't seem to be a thing to say.

"Don't you have view graphs to build?"

As Stretch walked down the passageway to the ready room, he reflected it was easy to call Little Lord, Skipper. With Fuller, however, he'd refused to use the term. The term *Skipper* connoted respect.

As Stretch reached for the doorknob, RT pushed open the door and stepped out of the ready room and stared.

The grin Stretch hadn't realized he'd worn got the heck off his face.

"Stretch," RT said. "I have to do a performance appraisal on you before I leave next week. I was at a loss as to what to say, but I got it now. 'Easily amused.' Has a ring to it, doesn't it?"

The following Monday, Helen Fischer hosted a luncheon for the new CO's and XO's wives, Sarah Fant and Laura Davison, in her on-base house.

Teresa was about to gather up the children to drop them by the childcare center when the phone rang. Helen wanted to know what Teresa was wearing.

"Maternity top and slacks. Little Pootzer won't squeeze into a dress anymore."

"Before you come over, dress down and tie a bandana around your hair."

"What?"

"Teresa, I don't have time to explain. Just do it, please, and invite Amy to do the same. Please?"

Teresa hung up and walked next door. When Amy answered the door, Teresa felt her mouth drop open. Amy's outfit was … perfect. Elegant but simple. Just a black skirt, ivory-colored blouse that looked tailored, rather than like a tent with enough material to accommodate the entire pregnancy, a baby-blue sweater.

"I have a pink sweater too. Mike wants a boy, and he insisted I wear the blue this morning. He's sure that will encourage a favorable outcome. Men!" Amy said.

Teresa told her Helen's message.

"Dress down?" Amy frowned. "This was going to be our coming out, the baby and me. Besides being a welcome to the CO and XO wives, of course. Did Helen say why?"

"No, but she did say please."

"Well, Helen has a reason, and that's enough for me. And fortunately, I have a bandana from a costume party Mike and I attended. Actually, I have two. Would you like to borrow one?"

"Yes, please. I was just going to tie my hair up in a scarf."

"That's a wonderful idea, Teresa."

"What?"

"I'm going to leave my outfit on and tie my hair up in a bandana."

Amy was ready to go, except for the bandana, which the ladies futzed with to get it tied on right. Then she rode with Teresa to drop off the children at childcare.

Back in the car and headed for Helen's, Teresa asked her about leave with their families back in Iowa.

"We had a wonderful time. Mike was a pop-his-buttons proud papa. Me, I was the baby's mother. That's how he talked about me the whole time we were home. Oh. You were gone when we arrived back on base."

"We ate in town, Mexican food."

"Anyway," Amy said, "Mike called RT. RT said Jon was going to admiral's mast but got a commendation medal instead. How did that happen?"

They were approaching family housing with the stonework façade pronouncing, "Alvarez Village."

"That sign makes me angry sometimes," Amy said. "It's like the navy doesn't want us to have too much room for pleasantness in our lives. We have to be reminded that LTJG Alvarez was the first POW, and the North Vietnamese are not done capturing new ones."

Teresa took Amy's hand. "Father God, in heaven, please watch over LTJG Alvarez and all of them in those godawful prisons."

"And watch over their families," Amy appended.

Together, they amen-ed.

Teresa let go of her neighbor's hand. "I'll give you a short version of the mast-to-medal saga. The whole story would take a day."

After Teresa's short version, Amy summarized, "The

previous CO, CDR Fuller, didn't like Jon and blamed him for wasting government money, but actually, CDR Fuller wasted it. RT figured out what Fuller had done and showed the admiral that not only had Jon not done anything wrong, but he'd actually saved the program while Fuller was ruining it. That's what you said, right?"

Teresa nodded. "Thank You, Father God in heaven, for the blessing of Helen and RT." She parked the car.

They got out of the car. Amy said, "May I say, Mrs. Zachery, how dressed up you look below? Mike says a bandana is an Iowa farmer's combination sweat mop and snot rag, and that particular accessory certainly makes you look dressed down above."

"How nice of you, Mrs. Allison, but I would like to point out that no one pulls off *dress up* as well as you, and I never would have thought to accessorize with an Iowa farmer's snot rag. Shall we, Mrs. A?" Teresa stuck out her arm.

The two pregnant Warhorse wives marched up to the door, where Amy pressed the buzzer.

Helen answered the door wearing one of RT's blue-jeans shirts with the sleeves rolled up, a pair of RT's jeans with the pant legs rolled up, and a red bandanna knotted over her hair.

The look on Helen's face surprised Teresa, even though it was her normal look, and today, in the midst of dressing-down frivolity, she looked concerned, worried that if she allowed herself to have too much fun, she'd miss the fact that someone needed help. Then her welcome-to-my-home smile crinkled the worry to pieces.

She invited them into her living room. A tall, trim brunette rose from the sofa. She wore jeans and a sweatshirt and her hair in a ponytail.

Helen said, "Maryann Toliver, meet Teresa Zachery and Amy Allison. Their husbands are the only two officers left from last cruise. Maryann's husband, Simon, is the new operations officer. Their moving truck showed up this morning. Two doors away, and I shanghaied her away from unpacking boxes."

"I begged her to shanghai me," Maryann said.

Shortly, Sarah Fant and Laura Davison arrived.

Helen served her guests quiche. Her guests served up a steady stream of polite, light conversation revolving around children, past duty stations, and mutual acquaintances. When a particular line of chatter among the older ladies wallowed into exclusion of the two youngsters, Sarah Fant steered the talk back over inclusionary ground.

The Tolivers and the Davisons had been on the East Coast where the new Warhorse ops O and XO had been instructors in the A-4 training squadron based at Cecil Field near Jacksonville.

Maryann said, "Originally, we both had orders to the A-4 training squadron here at Lemoore. Then the navy decided *Solomons* and her squadrons would decommission after one last cruise, and the A-4 training squadron here shut down, but the East Coast one kept going. Now the East Coast one is shutting down because the last of the A-4 squadrons there have decommissioned. But, here, *Solomons* and the Warhorses get to make a third final deployment. And the navy is standing up the training squadron here again."

Laura Davison said, "Gives me a headache trying to follow that story. Be nice if the navy could figure out what it's doing instead of jerking people back and forth and from coast to coast."

"It's not the navy. It's our stupid government." Maryann again. "They can't figure out how to lose this war. We sure aren't trying to win it."

Chilly silence dropped onto Helen's dining room table. Teresa held her silverware still.

"Ladies," Sarah Fant said, "our husbands are going to be very busy right up until they deploy in January. We will have to depend on each other. The squadron will be absorbing a number of pilots fresh from the training command and fresh from the altar of matrimony. These new navy wives will depend on us. The Warhorse Wives' Group must be welcoming and nurturing. And when we are together, everything we say must serve that purpose." Sarah's eyes roamed down the table and up the other side. "And no other."

As they drove away from Helen's house, Teresa said, "Mrs. Fant said, 'We won't wear our husband's rank in the wives' group.'"

"Right," Amy said, "and then I thought, *That's just what she was doing*."

"Me too, but what she was really doing was orchestrating things to keep us included. Maryann

Toliver kept dominating the conversation with Laura Davison. Sarah jumped in and maneuvered things back to where we weren't just two better-seen-than-heard JO wives."

"Last cruise," Amy said, "AB's wife, Sybil, was pushier than Maryann. But the CO's wife let her get away with it because Sybil got things done."

"It would have been interesting to see how things would have worked out if Sybil and Sarah wound up in the same group."

"The feathers would fly, but I'd bet on Sarah."

As they exited Alvarez Village and headed for the childcare center, Teresa thought of Maryann Toliver and Sybil Clark and her mother. As far as pushy went, three peas from the same pod. She hadn't asked Mother to come out and help with Little Pootzer, but she intended to. And she dreaded doing so.

Chapter 16

The Warhorses' new maintenance officer, LCDR Mark Wakefield, checked aboard on a Monday.

On Tuesday, the skipper called a meeting with the XO, the ops O, the maintenance officer, RT, and LT Zachery, the weapons training officer in attendance.

The CO explained CAG's philosophy for training the airwing for the next deployment. "We will prepare for ops over North Vietnam. Major emphasis will be training to conduct strikes from Alpha strike formation."

To conduct an Alpha strike, the airwing would gaggle up twenty bombers in four-plane divisions. Each division would act like a single plane. Around the Alpha gaggle, they'd package SAM protection, which amounted to the Warhorse TIAS birds, and MiG protection, which, of course, would be the F-8 fighters from the airwing.

The skipper said, "For those of us who will fly as bombers, our job will be to get our bombs to the target. The job of the F-8s and the TIAS birds will be exactly the same as ours: make sure the we get our bombs to the target. When we are loaded down with bombs, we are as maneuverable as a dump truck. If MiGs and SAMs cause us to dump our bombs to avoid getting shot down, they have succeeded in their mission, and we have failed in ours."

At a nod, the ops O, LCDR Simon Toliver, Simp, took over. "We are not jumping directly into Alpha strike training. We will have a detailed training plan for each of our pilots, and we expect a bunch of them to check in over the next two weeks. We will document what we can expect from each pilot, from the skipper to the newest newbie in terms of individual performance. How well can the individual bomb? How well does he maneuver his aircraft? These will be documented over the next couple of weeks. Once we get the basics down, we will shift training emphasis from individual to formation tactics. By the time we go to sea on *Solomons,* we will be focusing heavily on Alpha strike tactics."

"One last thing," the skipper said. "Some of us have combat experience over the North. Some don't. For the time being, we're approaching training as if none of us have combat experience. We all have to prove ourselves."

That afternoon, Stretch rode home with RT.

"How'd your hop with the skipper go?" Stretch said.

"He wanted me to show him the formation you and I practiced last cruise. I showed him, and he picked it up pretty quickly. When we got back, he asked me how many guys got comfortable with flying that way. I told him one.

"I've flown it with Mike Allison. He handles it, and he flies it fully invested in making it work. The other guys, at least last cruise, kind of said, 'This is stupid, but if that's what you want, I'll do it, but I ain't gonna let you kill me.' Most of them worried so much about how much extra gas they burned they didn't do the job properly."

"But Mike does?"

"Mike does."

"Good."

They rode in silence for a spell.

"Today was your last day?" Jon said.

"Movers pack us up tomorrow, load the van the next day, and we head out the day after that."

RT had a job in New Jersey coaching high school kids in baseball, basketball, and football. During the next couple of years, he'd be completing the requirements to teach history. Helen was also a teacher.

"You're not joining the reserves, just quitting flying cold turkey?"

"For me, it's the only way to do it. I like flying, but it's not something I can do as a sidelight. Flying demands everything of me I have to give. If I don't give her everything, she'll kill me."

Tires hummed over the two-lane blacktop sliced through cottonfields and leading away from the airfield.

"Stretch, I've heard you walk up to a plane and say, 'Hello, 510.' You know what I'm talking about."

Stretch nodded. But when he thought about flying in combat, ahead of him, he saw only a bank of pea-soup fog. RT was like a radar seeing the way through clear of mountains.

"Stretch, I'm at peace with what I'm doing. It would mean something to me if you were at peace with it too."

Gonna have to think about that.

High pucker factor accompanied contemplating the next cruise and flying sorties over North Vietnam. If RT were with the squadron, the pucker factor would diminish. It was just how things stood.

"Part of the reason I'm at peace with me leaving the navy is that you know more about fighting SAMs than I do. You have a good training program for the guys who will fly the TIAS planes. Of course, you'll never be the stick-and-throttle jockey that I am, but you are halfway close to sort of good. So, I am at peace with leaving it all in your hands."

"My hands?"

"Yep."

"And I'm supposed to find peace in that notion?"

"Yep."

They drove past the Alvarez Village sign, and Stretch felt peace take a step backward.

RT pulled into the Zacherys' cul-de-sac and stopped. Jon thanked him for the ride and got out. RT put on a faint and lopsided grin and drove away.

Zachery watched him until his car disappeared behind the edge of his quarters.

Inside the house, two "Daddys!" and two *urk ahs* and dinner with the most beautiful pregnant woman in the universe slapped Band-Aids on his heart.

The next morning, Jon addressed an envelope to "Dear East Coast Puke." Inside, he put a note:

It took some doing, but I got there last night, all the way to peace.

Jon dropped the note through the mail slot in door of the Fischers' quarters. Then he went to work.

The next day, the Warhorse Wives' Group shanghaied Helen Fischer for a farewell luncheon hosted by Maryann Toliver. The wives all wore bandanas tied over their hair. RT, left to supervise the loading of boxes and furniture onto the moving van, also had a red bandana tied over his crew cut.

The Warhorse Wives' Group had grown to eleven members. Deborah Wakefield was the wife of the maintenance officer. Naomi Engel was married to Harvey, a senior LT and combat-experienced pilot. Tara Wisdom's husband, Howie, was also a senior lieutenant and a ground maintenance officer. All the others were married to pilots. Then, besides Teresa and Amy, Monica Newsome, Lydia Foster, and Wanda Mason were all married to junior-grade lieutenants. The Fosters checked into the squadron directly from their honeymoon.

At the luncheon, Wanda Mason sat on a sofa next to Amy Allison and Teresa. Wanda asked Amy, "Everyone has been so welcoming, and Sarah Fant treated me like … like I'm her equal or something. Is this normal?"

Amy smiled at Teresa and then answered, "It's normal in the Warhorse Wives' Group."

That night, Teresa told Jon about the wives and their *normal*. "Are you guys the same way in the ready room?"

When she thought he wasn't going to answer, he did. "Naomi Engel has three children. Her youngest daughter is the same age as Jennifer. You like Naomi."

"And you don't like Harvey?"

"It's not that I dislike him. He thinks I had one hop over North Vietnam, and he's had a hundred. What could I know that he doesn't? That's the way he looks at it."

Teresa said, "If I had my way, Jon Zachery, I would have you be happy with your squadron mates rather than me happy with the wives' group."

"It's not that big a deal. There's always some friction. When I was on the destroyer, I remember Admiral Ensign and Dormant. One thought he was better than the rest of us JOs. The other didn't care who was better; he just didn't want anything to do with the navy or the rest of us. In the Warhorses' last deployment, we had Amos Kane."

"The way Sarah Fant brought all of us into line, to be supportive and inclusive, I thought was masterful. Why can't the skipper do the same thing with you guys?"

"He has laid out how we are going to train and operate. Nobody argues with him. There're just a few sparks among a few of us menials. I don't think the skipper thinks that's bad. The navy values competitiveness, aggressiveness, the will to win. If you cram a bunch of those types together, there's going to be snapping of teeth and snarling. As far as CDR Fant goes, he's a good CO. I think as highly of him as I do of RT."

"I just worry about you, Jon Zachery."

"I worry about you more."

"No. I worry most."

"So, it seems you qualify to be a naval aviator, Teresa Velmer Zachery. On the basis of competitiveness, only we are out of maternity G-suits."

Then the Robsey twins checked in to the squadron. Both of them had black hair and were five nine or ten. Lieutenants Rob White and Robert Stoll didn't look like twins. Sharp planes comprised Rob White's pale face, a sharp, pointy nose, slab cheeks meeting at his small chin. Stoll's puffy face had a hotdog-bun nose sitting in the middle of it, and big ears framed it all.

Mike Allison had named LTJG Nat Newsome Nose. He was the personnel officer in the squadron, and on Friday afternoons, he went to the enlisted club with his sailors and bought them beer. After a brew or two, they gave Nat the skinny, the poop, the word. One Friday, one PN (personnel man) disclosed that he had a sister who worked for CDR Fuller's father-in-law. The father-in-law was the CEO of an East Coast insurance company and he gave Fuller a job. According to the

PN's sister, father-in-law gave Fuller the job so that his son-in-law wouldn't move in with them.

When Nat Newsome shared that tidbit with the Warhorse JOs, Mike gave him his call sign. Nat Newsome had some kind of Nose for news.

Shortly after that, Nose gave himself responsibility for giving new pilots their call signs.

Newlywed Terry Foster became Nooner because he went home for lunch.

LTJG Oliver Mason got his call sign the first time he stood SDO (squadron duty officer) watch. The SDO did not write the flight schedule, but it was his job to manage it as a direct representative of the CO. Stretch was on the schedule as lead of a four-plane. Harvey Engel was two. Alice was three, and Nooner was four. Everyone sat in the circle of ready room chairs for the brief, except there was no Harvey. At 0745, Harvey still had not shown up, but Robert Stoll came in for a cup of coffee. Oliver Mason, as SDO, added Stoll in Engel's place, and Stretch went down the preflight briefing checklist. At 0800, Harvey Engel showed up and wanted Stoll to get the hell out of his seat for the brief.

Stretch said, "You're a half hour late. I already covered most of the brief. I'm not starting over just because you can't make it here on time."

Harvey stomped to the SDO desk. "Tell that little shit Zachery to put me on the flight."

Oliver said, "You're a half hour late. Stretch already covered a lot of the brief. You didn't call to let me know you were going to be late. You're off the schedule."

Engel shook his finger in Oliver's face. "You no-account dipshit JG newbie, I'm a department head. Put me back on the schedule. That's an order."

Oliver jumped to his feet and lifted up the three-ring binder with SOP (standard operating procedure) on the cover. "This says the SDO is the direct representative of the commanding officer. As his rep, I'm telling you you're off the schedule. Furthermore, I'm ordering you to knock off the shouting. There's a four-plane about ready to man up. They should have their focus on flying, not have to listen to your juvenile hissy fit."

Harvey stomped out and, a minute later, returned with the ops O.

Simon Tolliver ambled up to the SDO. "Um. Now see here, Oliver. Harvey is a department head. He's got, you know, important stuff to do. A few minutes late for brief is no big deal."

"Ops O, he wasn't a few minutes late. He was thirty minutes late. He missed the brief. I'm not putting him on the schedule."

"What'd I tell you!" Harvey shouted.

The skipper walked in. "What the hell is going on here?"

"Mason jerked me off the schedule, Skipper," Harvey said. "Simon and me, we were trying to tell him to put me back on, but the arrogant little prick won't do it."

The skipper asked Oliver for his side of the story. After he heard it, the CO shook his head. "Harvey, Simon, you're my department heads. As such, *your* job is to make *my job* easier, not more difficult. Get the hell in my office now." The skipper turned to Stretch. "You ready to go?"

"Yes, sir. We need to man up, or we'll miss our scheduled time on the bombing range."

Stretch led his flight to a bombing range at Fallon, Nevada. Each plane carried six practice bombs. Each pilot recorded his first bomb hit and his average of four bombs dropped in manual mode. Each pilot released his last two bombs using the CP-700 computer.

After the flight, Stretch updated the stats on pilot bombing scores. Alice and the skipper were the best bombers based on first bomb and average of bombs dropped. Stretch and Stoll scored second best. The bomb scores were recorded on the whiteboard in front of the ready room.

In addition, Nose commandeered a corner of the whiteboard and wrote down pilot call signs as he came up with them. Simon Toliver, the ops O, became Simp, for Simple Simon. Simp had a slow-talking, "aw shucks, man" demeanor. Mark Wakefield became ECP for East Coast Puke, but Nose shortened it to EC. Everybody would know Puke was implied.

Harvey Engel became Not. Engel meant *angel* in German, and Harvey was not an angel.

Rob White and Robert Stoll became the Robsey twins, with LT White earning Blackey as his call sign. Rob White spewed vulgarity and profanity every time he opened his mouth. Pure white he was not; rather he was black as sin. Stoll became Troll because he looked like one.

A TICKET TO HELL: ON OTHER MEN'S SINS

Oliver Mason became Skippy, because he'd behaved like a little skipper on his first SDO watch.

Blackey took over the best bomber spot on the whiteboard, and the CP-700 computer occupied the cellar. The CP-700 bombing computer proved to be reliable half the time. When it worked, it bombed better than Blackey, but 50 percent of the time, it lobbed bombs a thousand feet, or more, long.

The other item recorded on the whiteboard was a listing of pilots qualified to shoot Shrikes for the anti-SAM mission. The CO and XO would only fly as bombers. The other pilots, except for Not and Blackey, were qualified.

The skipper called Stretch into his office. "What about Not and Blackey? Why haven't you designated them to shoot Shrikes?"

"Those two think they know more about flying over the North than I do. They won't take the drills I run the guys through seriously. If we are over the North, and I have a TIAS bird and they have a regular A-4, I want to be able to tell them turn left ten degrees, pull up ten degrees, and shoot. I'm not sure I can count on them to do what I tell them. I don't think they'll listen to Alice either."

"How about if I talk to them?"

"If you do, they'll go through the motions, but we won't be able to count on them to do the right thing in the heat of a SAM encounter. I'll bring them around."

"Stretch, I don't want to fool around with this. We don't have time for this independent cowboy crap. I'm going to help you."

Chapter 17

Teresa pulled into an open parking spot adjacent to the Warhorse section of the hangar, turned off the engine, and checked on the children in the back seat.

Jennifer colored. EJ flipped the cardboard pages of his book while he sucked the other thumb.

Jon and Mike Allison exited the hangar and headed for the car. Teresa scooted to the center of the front seat.

Jon got behind the wheel and said, "Well?"

Teresa'd just come from her twenty-four-week checkup. "All is well. With Little Pootzer and me."

Mike had taken the shotgun seat. "Great."

"Amy told me she and the baby got a clean bill of health on her checkup this morning," Teresa said.

Mike grinned as if he'd done something good.

Jon squeezed Teresa's hand. "Thank You, God."

Jon helloed the back seat, and the children echoed back with "Hi, Daddy"; then he backed out of the parking place and got the car headed for home.

Mike looked past Teresa. "D'ja meet the Wiz?"

He'd asked the question of Jon, but Teresa cut in. "Who's Whiz?"

"Our assistant maintenance officer, a ground pounder, Howie Wisdom. He checked in today."

"So Whiz, W-H-I-Z?"

"Well, uh, no," Mike said. "W-I-Z, as in U-R-I-N-E."

From the backseat Jennifer asked, "What does U-R-I-N-E spell?"

"Nothing," Jon said. "I'll explain when you're ... twenty."

Teresa said, "Pooh. Pilot call signs have to be embarrassing, profane, or racy. Sort of juvenile, don't you think?"

"Not sort of juvenile, Teresa," Mike replied. "Very juvenile. That's a requirement too. And Nose is a good call sign picker. As good as Tuesday was last year."

Teresa glanced at Jon. He gripped the wheel with both hands and stared straight ahead. His jaw worked as if he was grinding roughage with his molars.

Tuesday.

Sometime near the end of last deployment, Jon and Tuesday had had a falling out. Jon wouldn't talk about it, but it was deep-seated. Mention Tuesday, and it was as if Jon flinched.

Teresa changed the subject. "I'm looking forward to the party at the skipper's house tomorrow. We get to meet Carolyn White."

Jon and Mike exchanged meaning-loaded glances.

"What?" Teresa asked.

Jon said, "Later."

Quite a bit later, after the children were in bed, Teresa remarked, "It's later."

"You want to know why Mike and I looked at each other the way we did?"

"Precisely."

"Okay, but first, you said all is well with Little Pootzer and you, but that isn't the whole story, is it?"

"No. My doctor wants to see me once a week from now on."

"He's worried?"

"Based on what happened with Daniel at twenty-eight weeks, he just wants to keep closer tabs on us. So, it's not something that should cause you to worry more. He's being cautious. You should worry less."

"Sometimes, Teresa Velmer Zachery, a thing can make all the sense in the world, but it still doesn't amount to a logical hill of beans."

"Jon—"

"God won't give us anything we can't handle. I know. But I worry about what's coming up. Jennifer starts kindergarten this month. And October, November, and December, we're gone more than we're home. My trust in God is complete but completely drenched in worry."

"There's one sure way to wring worry out of a worried soul." Teresa pulled her rosary out of her pocket and started to get up to kneel by the sofa.

"You stay right there, Teresa Velmer Zachery. I will kneel for the three of us."

They said the prayers, then got ready for bed. "Okay, Jon. Now tell me what you and Mike were thinking when I brought up Carolyn White's name."

Carolyn White worked as an airline stewardess based out of Los Angeles. She had an apartment there.

Sarah Fant said Carolyn had worked her schedule so she'd be able to spend ten days a month with her husband in Lemoore. The wives were all anxious to meet her.

Teresa waited a moment. "Are you thinking about how to tell me, or are you not going to answer?"

Jon had a number of things he wouldn't tell Teresa. Just then, he didn't want to add another to the list.

"This afternoon, you said pilot call signs have to be embarrassing, profane, racy—"

"And juvenile."

"That too. There's a short answer to your question, but will you let me give you the long version?"

"Either one."

"RT told me once that a good pilot pushes the envelope—that he flies in the gray area between safety and recklessness. 'A good pilot knows exactly where recklessness starts, and he flies right up to it.'"

"Jon Zachery, do you think I don't worry enough about you and your flying?"

"You worry exactly enough. But God put me together with RT so I could learn from him, and I have. But with this pushing the boundary thing, pilots get the sense they can do things earthbound mortals can't. So, we act that way on the ground. We push the social boundaries the same as we push the flying boundaries, because those things just don't apply to us immortal, invincible naval aviators the way they do to the aforementioned earthbound mortals."

"You can't be talking about the man I'm married to."

"Teresa. When we're together in a pack, we operate to our own standard of acceptable behavior, and it has to be a little outrageous."

"A little outrageous? That's an oxymoron."

"For everyone else maybe but not for naval aviators. Our behavior needs to be close to outrageous but not actually there. And whether I engage in the behavior or not, I accept the pack behavior as … acceptable."

"I still don't think you act that way."

Jon shrugged. "In the Warhorses, we had our way of behaving. Then Blackey showed up. He's *the* best pilot. Bombs best. Flies defensive maneuvers against enemy fighters best."

Teresa waited. She knew more was coming.

"On the ground, to Blackey, it's as if the boundaries that apply to the rest of us don't apply to him. He has to be more profane, more embarrassing, more juvenile than the rest of us. It's sort of like he has to remind us constantly that he's better than us, that he can get away with things we can't."

"Okay?"

"Today, Rob started flashing a polaroid of Carolyn in the nude."

"What! She's coming to a party tomorrow night, and he's showing you guys pictures of her naked!"

"That's Blackey."

"Why does the skipper let him get away with that kind of thing?"

"I doubt if the skipper knows. Blackey is *probably*

smart enough to know the CO would smack him down for that. So he only showed it to us JOs."

"Poor Carolyn."

The next morning, Stretch was scheduled to fly with Not, Alice, and Blackey for an Ironhand training mission. Ironhand was the navy name for the anti-SAM mission. By seniority, Not would have been lead, Alice two, Blackey three, and Stretch four. But the flight schedule listed them as Zachery, White, Allison, and Engel.

The skipper, Stretch, and Alice were in the ready room at briefing time. One minute late, Not and Blackey walked in talking and laughing. They saw the CO sitting in the circle of chairs for the preflight brief and hustled over.

The skipper looked at his watch. "Don't sit down. Go to the grease board and draw the Shrike envelope for the case where the shooter is at ten thousand feet, speed Mach .7. Both of you, draw the envelope. And show five, ten, and fifteen miles from the target SAM site."

Not drew something that Edgar Jon might have drawn.

Blackey said, "I don't have the envelope memorized, Skipper."

Skippy was SDO, and the skipper invited him to draw the envelope, which he did quickly. "How'd he do, Stretch?"

"Right on, Skipper."

"How'd the two combat-experienced veterans do?"

"Not probably looked at the envelope in the weapons manual once or twice, but Blackey doesn't have a clue."

"You two dimwits are combat-experienced senior lieutenants. You should be flying as Ironhand leads. As it is, you are the only two guys Stretch hasn't even qualified to be Ironhand wingmen."

The skipper locked eyes with each of the dimwits. "If you do not qualify as Ironhand wingmen, I do not need you or want you in the Warhorses. If it comes to that, I'm not going to transfer you to another squadron in the airwing. I'm going to put in your fitness reports that you are not fit for flying combat missions and should be assigned to nonflying jobs to serve out your time."

The skipper focused on Not. "I've told you this before. You're one of my department heads. Your job is to make my job easier, not harder. This is the last time I'm telling you that. I could put Skippy in as admin officer, and I'd bet money he'd do the job better than you because he takes things seriously."

The skipper shifted his aim to Blackey. "You think you're better than anybody else because you bomb better, but you don't know your ass from a hole in the ground when it comes to the Ironhand mission. And to our airwing, Ironhand is our reason for existing."

The skipper flicked his eyes back and forth from Not to Blackey. "This is how it's going down. I'm flying with you guys this morning to observe, and if you don't fly the mission exactly as Stretch briefs it, when we come back, you will pack your gear and get the hell out of my squadron. In the unlikely event you do fly the mission properly, Monday morning, Not and Blackey, you will prove to me and Stretch you know as much as Skippy about the Ironhand mission. Or you will pack your bags and get the hell out of my squadron. Now, Stretch, brief the hop, and let's go aviate."

The flight was a combination bombing hop to a scored range and an Ironhand training mission on a new simulated SAM site at Fallon, Nevada. All four pilots flew TIAS birds. The simulated SAM was designed by Ross Hill and built by technicians at China Lake.

"You'll do two kinds of runs," Skippy explained. "On one, it will be as if the SAM site is targeting you. On the other, it will be as if the SAM is targeting someone else. The key is TIAS is a very sensitive system. It can detect a SAM radar even if it isn't targeting you."

"Thanks, Skippy. Go back to your SDO desk." The skipper faced Blackey. "What's the significance of what Skippy said?"

Blackey's face remained set in a poker-player blank, but his eyes blinked rapidly. "Um, a TIAS bird and a wingman could prowl to the side of an Alpha strike ingress route and see the SAM that targeted planes in the strike and kill the site without the North Vietnamese gomers knowing the Ironhand guys were coming."

The skipper grinned. "That's excellent, Blackey. See, you are teachable even from the junior pilot in the squadron. All you have to do is shut your arrogant mouth and pay attention. What else, Stretch?"

Stretch completed the preflight briefing checklist, then they manned up, flew the mission, and returned to Lemoore. In postflight brief, Stretch asked Not and

Blackey to recount what they learned flying against the simulated SAM.

After they recounted their experience, the skipper dismissed them, and they entered the locker room. "They paid attention. What do you think, Stretch?"

"The issue is trust, Skipper. Everybody else in the squadron gets it, and I trust them to fly the mission. They will follow their lead's orders promptly and exactly. These two guys flew the mission as briefed because you were watching. I don't trust them. Yet. I want to fly with both of them next week, and we'll see how they do without you watching."

"Good," the CO said. "See you at the party tonight. I'm looking forward to meeting the woman who married Blackey."

"Everybody is, Skipper."

Chapter 18

The party started at 1700. Jon figured they'd spend two hours there, and then he'd get Teresa home.

By the other cars parked on the street, they were the last to arrive. A note posted by the front door read, Come On In. When he opened the door, a flood of happy noise full-immersion baptized them.

The skipper's wife took Teresa's arm and led her to a sofa in the living room. Amy sat on one end of it. Opposite her, Blackey and a brunette occupied overstuffed chairs. Sarah Fant introduced the Zacherys to Carolyn White.

Even sitting, Jon could tell she'd be taller than Blackey. She was beautiful. Jon was hard-pressed to say whether blonde Amy Allison or Carolyn was the most beautiful. Both women were, as the saying went, very nicely put together.

Jon told Carolyn he was very pleased to meet her, but he was thinking, *Teresa Velmer Zachery, in the beauty department, you ain't no slouch either.*

It was interesting, seeing slender Carolyn opposite the two pregnant Warhorse wives. The purpose for attractiveness in women reflected in the three most beautiful at the party. When they returned home, he would tell Teresa his observation.

Blackey asked Amy and Teresa if he could get them a drink. Jon offered to go with him.

"No, no," Blackey said. "Stay and entertain the ladies."

Carolyn said, "Teresa, Amy told me she and Mike met in college. How did you and Jon meet?"

"High school. We started going steady in our senior year, and we still are."

Carolyn's face lit up. "That's so sweet."

Jon was sure she'd sincerely meant it. He asked her how she and Blackey met.

"First of all, this pilot call sign business, I am not used to it yet. But I met Rob, Blackey, on a blind date. My father works for the King Ranch in Texas. I was home to visit my parents, and one of my friends from college invited me to accompany her to a party at the Officers' Club at Naval Air Station Kingsville. Rob/Blackey"—Carolyn smiled—"You all have gotten my husband stuck in my head as that way. Anyway, he made me laugh. I felt like I hadn't ever been really happy before I met him. I fell in love with him that first night. A few weeks later, Rob took leave and visited me in LA, and we got married. That was six months ago. In that time, we've had maybe three weeks together. I am so happy to have him on the West Coast now."

Talking with Carolyn was pleasant. With Blackey, a perpetual air of unpleasantness hung over every conversation Jon'd had with the guy. Maybe with Carolyn around more, she would civilize him.

Blackey returned with sodas for Amy and Teresa. Troll was with him carrying a drink in each hand. Sarah Fant invited Carolyn to come to the kitchen and meet some more of the ladies.

The Robsey twins went back out to the patio through the open sliding glass door. Most of the men and a bar were out there. Mike Allison entered from the patio and handed Jon a Jack Daniels on the rocks.

Jon nodded thanks. "Did you see what Blackey did?"

Mike shrugged. "I'm not sure my brain believes what my eyes said just happened. Blackey acting nice! You think he just missed his wife?"

Stretch shrugged.

Not stepped in from the patio. "Stretch, can I talk to you a minute?"

Jon excused himself.

"Blackey talked to the XO about rooming with Troll on the carrier," Not said. "The XO okayed it. That means you and me are going to be roomies." Not grinned.

Blackey came up to them and handed Stretch a fresh drink and took his barely touched one away.

"Here's the deal, roomie. Blackey and me, we were wondering if you'd come to the squadron tomorrow morning and help us get up to speed on the Shrike and Ironhand stuff? We were thinking a reasonable time, like ten."

Stretch walked to the edge of the patio and dumped his drink behind a shrub; then he returned to Lieutenants

Engel and White. Both were senior to him. "If you want help with the anti-SAM mission, I will be at the squadron at seven tomorrow morning. If you are one second late, I'm leaving, and you're on your own."

"Jesus Christ, Stretch. It's Saturday tomorrow. I haven't seen Carolyn for three weeks."

"The time is now six forty-five."

Blackey snarled, "You son of a bitch."

"Six thirty."

Not grabbed Blackey's arm and pulled him away.

The skipper left the bar and walked over. "Everything okay, Stretch?"

His anger hadn't coasted down much. Stretch took a deep breath and huffed it out. He smiled. "A-Okay, sir. And, uh, by the way, great party."

The skipper sniffed. "Do you smell that?"

"Smell what, sir?"

"Mendacity. But I smell something else. Food's ready. Come on."

The Fants' dining room table accommodated twelve. Ten of the wives sat there while Sarah Fant and Maryann Toliver worked in the kitchen plating food for the men, who returned to their haven in the backyard.

Not pulled a lawn chair next to Stretch's. "Look. I appreciate you coming out to the squadron for Shrike stupid study. I should have paid attention when you briefed everyone else, but I didn't. So, I'm sorry for screwing up your Saturday morning, but I'll be there at six thirty."

Stretch chewed and swallowed . "You keep this up, Not, and Nose is going to change your call sign to Yes-He-Is-One."

"Oh, I doubt that. Besides, I kind of like Not. Though of course, I know it'd be a mistake to let Nose know that. It's a sure way to get one I won't like."

"Your secret is safe with me."

Stretch and Not talked about their children. Then Not related a bit of personal history. In 1968, after serving his obligated time, he left the navy and landed a job with an airline. He completed training and was beginning to get himself established when he was furloughed. With no other good options in front of him, he came back in the service with his previous rank and seniority.

Stretch related his history. When it was time for dessert, he mused he'd had a pleasant chat, an unexpected, pleasant chat, and further mused he'd have such a conversation with Blackey right after hell froze over.

After dessert, Sarah Fant presented a Warhorse Wives' Group pin to Carolyn White and welcomed her to the group. Shortly after that, the Zacherys left for home.

The next morning after Shrike stupid study, Blackey bolted out of the ready room. Stretch and Not walked to the parking lot together.

"Did you hear what happened with Carolyn White last night after you left?"

Stretch shook his head.

"Carolyn found out Blackey had shown the nudie

of her to the JOs. Apparently, Troll told her. She left the party with Troll, without telling Blackey, and drove back to LA. I think Blackey is driving down there to see if she'll come back with him."

On Sunday, Blackey drove Carolyn back to Lemoore. On Monday, Not and Blackey passed their test with the skipper.

Also on Monday, Sarah Fant hosted an emergency luncheon to reassure Carolyn White she was welcome. Maryann Toliver informed Carolyn that being married to any man was enough of a challenge for most women, but only the toughest could handle being married to a naval aviator. And Maryann presented her with a needlepoint that proclaimed: Toughest Job in the Navy: Navy Wife.

Teresa told Jon she'd never been to an *emergency luncheon* before. "Carolyn is a wonderful person. All of us like her, and all of us are glad she came back."

Tuesday morning, the XO called Jon to his office. "Blackey and Troll were going to room together on the ship but not now."

"I heard Troll told Carolyn about the nude photo. Why did he do that? He and Blackey were such good buds."

"At the party, Carolyn went out to the bar on the patio, looking for a glass of wine. When she walked out, Nooner, Nose, and Skippy were huddled in a circle. They were talking about Carolyn, how she seemed like a really nice lady, and they were wondering how Blackey could do such a thing to her. When Carolyn showed up, they broke apart. When Carolyn went back inside, the first person she ran into was Troll. She told him how the three JOs acted when she walked in on them. Troll didn't want to tell her, but she knew he was hiding something. She pressed him until he told her."

"And now Blackey blames Troll."

"Right. And that's why you're here. You and Blackey are going to be roommates on the ship."

"What? I thought I was rooming with Not."

The XO shook his head. "Not and Wiz are both senior to you, and they don't want to room with Blackey. That leaves you. Congratulations."

"Crap!"

"The proper response, Lieutenant, is 'Crap, aye, aye, XO.'"

Lieutenant Zachery popped to attention. "Aye, aye, XO." Then he did a left face and marched out of the office and through admin muttering, "Crap, crap, crap," as if he were counting cadence.

Through the week, Stretch flew with Not and Blackey twice each. He qualified both of them as Ironhand wingmen, but only Not was designated as an Ironhand lead.

The following Monday, the skipper met with the ops O and Stretch.

"During the upcoming at-sea period, CAG will have us doing a number of practice Alpha strikes. You have the number of Ironhand leaders you need, Stretch?"

"Yes, sir. LCDR Wakefield, Not, myself, Troll, and Alice. We can cover three Alpha strikes a day."

"Okay," Simp said. "We have three new JGs checking in today. Get them as much Shrike and TIAS training as you can this week."

From the CO, "You're still not ready to qualify Blackey as a lead?"

"It's not a matter of ability. Once you got his attention, he buckled down and learned how to do the mission. It's a matter of trust. If it's up to me—"

"It is."

"Then, no. And unless something extraordinary happens, I only want him to fly as my wingman."

The next day, Jon and the squadron would fly out to the *Solomons* for three weeks of training. That night, Jon knelt beside the bed as they prayed a rosary. Jon used "Please, God, watch over Teresa and Little Pootzer," mysteries. Then he climbed into bed.

"I talked to Mother again today. She said she will come out whenever we need her and will stay as long as she has to."

Jon suppressed a sigh. His mother-in-law spoke an eloquent body language. She never said so verbally, but she'd never approved of him. When they dated in high school, she tolerated him, but she conveyed an air of *Teresa, you can do so much better than this*. After high school, he entered the navy, and Teresa started school to become an RN. Mrs. Velmer brightened, convinced,

Jon was sure, the romance would fritter itself out and that Teresa would snag a doctor to marry. Except it hadn't happened that way.

Jon would never forget their wedding reception. At the point where Teresa danced with her father, and Jon should have danced with his mother, only his mother did not and would not dance, he'd partnered with his mother-in-law. It had been like holding an iceberg.

Teresa squeezed his hand. "When we've needed her, she's always come to help."

"Yes, and thank You, God, and thank you, Mrs. Velmer."

His mother-in-law was even more an alpha female than Maryann Toliver. The world could not get along without her; that, Jon thought, was Mrs. Velmer's motivation for helping them. But he kept that subjective analytical tidbit to himself.

"Jon, you're flying to the carrier tomorrow. You worry about that. Little Pootzer and I will be fine. Sarah Fant said she is going to check on me every day."

"I just wish I could be here when you have your twenty-eight-week checkup."

"The doctor doesn't need to examine you, silly. We'll be fine. And Sarah has spoken to the base chaplain. He'll give you daily sitreps. Sitrep. See what you've done, Jon Zachery. You've turned me into a regular navy wife. Now come here."

"Uh—"

"This is not like the night we lost Daniel. In fact, he's here with us, and it makes him happy to see how much in love his parents are. Do you feel him here?"

Which was impossible to answer with the heavy-duty lickey-face going on.

The next night, Jon was in the stateroom—the same one he'd occupied returning from last cruise—writing to Teresa. He'd completed his night carrier landing re-qual. Blackey was flying. The noises of cat shots and landings and of planes moving from the flight deck to hangar were the welcome-home sounds with which a carrier greeted pilots back aboard. There was comfort of a sort in those thumps and thuds.

Each thump and thud, though, bored the hole in his heart a little bigger. He was only a couple of hundred miles away. A half hour flight in his A-4, except he couldn't fly there. It was the same as being ten time zones away from Teresa and his family.

He'd just come from the chapel, where he'd sat in the back row of chairs and prayed a rosary into the sacred silence for Teresa and Little Pootzer, and Jennifer and EJ.

Now he sat in the bubble of light from his desk lamp with half a page filled with lines, and writing drew him into his writing-a-letter-to-Teresa world. The thumps came from farther, and farther still, away.

The door to the room opened.

"Jesus—F-word—Christ. Why the—F-word—you sitting in the dark?" Blackey turned on the overhead light and walked to his fold down and looked at the letter.

"Dearest Teresa. What the hell? We've been gone for twelve—F-word—hours and you're writing her an—F-word—letter!"

Jon stared at his pad of paper as anger mounted and hesitated just short of rage.

"I bet you're telling her how much you miss her big pregnant tits."

Jon bolted up and backhanded Blackey across his mouth. Blackey held his hands over his bloody lips. Jon punched him hard on the chest, and he flew back against the door and slid to a sitting position.

Jon grabbed handfuls of flight suit, jerked him to his feet, got in his face, and snarled through clenched teeth. "There are going to be rules in this room."

"I'm not following your—F-word—"

Jon slapped him four times. Blackey never even raised a hand to try to defend himself.

Jon stared into Blackey's dull eyes. "There are going to be rules." Jon jabbed a finger into his chest, and Blackey flinched. "One: no cussing. None." Another finger jab. "Two: you do not ever let a word about Teresa escape from your filthy mouth."

Jon patted Blackey's cheek, and then Jon looked at his open hand as it clenched into a fist. Blackey's face showed nothing.

He'd make a heck of a poker player. But he wasn't a fighter, which was strange. How could he be such a big horse's butt and not know how to defend himself?

Jon locked his letter in his safe and went back to the chapel.

Chapter 19

Seven months pregnant Teresa parked her car in the carport, walked to her neighbor's back door, and knocked. Seven months pregnant Amy Allison answered the knock.

"Teresa, what's wrong?"

"I have to go in the hospital, Amy. To be safe, my doctor wants me confined to bed for the next month. If I hang onto Little Pootzer until I get to thirty-two weeks, he'll let me come home."

"Oh Lord, isn't that the way of it. The guys leave for workups, and the next day, things like this happen." Amy stepped out and moved to embrace her. Their tummies made it awkward. "What can I do to help?"

"Jennifer and EJ, I need to figure out who will take care of them. I have to pack and be at the hospital by one. Three hours. I'll call my mother and see if she can come out."

"Teresa, you call your mother. I'll call the skipper's wife. Then I'll come over, and we'll get the rest of this organized."

As Teresa dialed, she wasn't worried about her

mother's reaction. When she needed her, Mother came to help. As she had with Jennifer and Daniel. Father, however, after Jennifer came into the world via emergency C-section, told Jon he should not endanger Teresa's life by getting her pregnant again. Jon believed her father. Teresa, however, would not agree to birth control, morally wrong according to the church.

"That leaves celibacy," Jon said. "Priests practice celibacy, but they don't sleep in the same bed with the most desirable woman in the universe. I'll sleep in Jennifer's room, and she can sleep with you."

"We will not live that way," she'd told him. "I am the one facing the surgery. It surprised us this first time, and I won't lie. It scared me. But now I know what it's like, and I'm not afraid of it. The Lord will not give us more than we can handle."

His faith, she knew, fell short of her rock of faith. If something happened to her, he would blame himself for giving in to lust. She'd told him their marriage was always a sacrament. "Always, and never a deadly sin."

After they announced her second pregnancy, her father wouldn't speak to Jon ... until the birth of his grandson. Then he acted as if Jon had done the most wonderful thing, as if Teresa had had a minor part in it. For him, a grandson made all the difference. When Daniel came into the world, another male, losing him made the situation a tragedy that the whole family shared in. But what would he say now with another complicated pregnancy to deal with?

Men!

Her mother picked up. "Of course, I'll come out,

dear. I'll get there as soon as I can." Mrs. Velmer paused. "How will I get in touch with you? Who will watch the children until I get there? Will somebody be able to meet me at the airport?"

Teresa gave her Amy Allison's phone number and assured her that Amy would call her that evening when a few things had been worked out. "The squadron wives will pitch in and take care of things until you arrive." Another of her articles of faith.

Teresa hung up the kitchen wall-mounted phone next to the bulletin board. She looked at the calendar with days marked when Jon would be gone. The calendar would go with her to the hospital.

Amy knocked and entered.

"I'm so glad Sarah Fant is here and we don't have to deal with Mrs. Fuller," Amy said. "She's calling all the wives. Everybody will help. Once we have you settled in the hospital, I'll call Sarah. You're not to worry about Jennifer and EJ. Worry about yourself and the little one."

The two women embraced again.

"I need to call Naomi," Teresa said.

She had dropped the children off at Naomi Engel's house that morning. Engel meant *angel* in German, Jon told her, and his call sign was Not. Harvey might not be an angel, but Naomi was one.

"I called her," Amy said. "She's bringing the children over. After you pack, we'll have lunch together. Then I'll take you to the hospital. Naomi will take care of Jennifer and EJ as long as you need her to."

"That's a lot to put on her when she has three of her own."

"Naomi says, 'You have a child, he or she fills your life. Have two kids, they fill your life. Have twenty kids ... see where this is going?'"

In the Warhorse ready room, call signs sewn into covers over chair backs identified each pilots seat. The chairs were then arranged according to rank. In the second row, behind the CO and XO, Stretch's chair sat next to Blackey's chair. Both hosted namesake occupants. Both pilots waited for their time to man up for their night flight.

He used the time to write to Teresa.

Blackey humphed—at him writing to Teresa, Jon presumed.

"Holy shit!" Blackey said.

A thump shuddered through the hull.

The ship's announcing system blared, "Fire, fire, fire on the flight deck."

Blackey said, "An F-8. Classic. He went high, pulled power, fell like a rock, and smacked the ramp. Poor son of a bitch."

Jon looked at the closed-circuit TV screen in the front of the ready room that showed the landing area of the carrier. The fire whited out the screen. This set of workups was starting the same as the previous one. A year and a half ago, during carrier quals, an F-8 flew into the steel cliff at the rear of the ship. Now, as then, he thought of his friend Tiny, an F-8 pilot. Tiny was a great stick and throttle jockey. It couldn't be him.

Please, God.

A sailor wearing a sound-powered phone rig occupied the chair against the bulkhead in the CO's row. Flight deck control passed information to all the ready rooms via the phone system.

Sound-powered-phone Sailor reported, "It was a JG from VF-12."

Tiny was in the other F-8 squadron, VF-14.

Thanks.

Oh, and that JG's family, well, You know.

"You're just going back to your letter?" Blackey shook his head. "An F-8 guy gets snuffed, and it's no big deal?"

Blackey got up, poured himself a mug of coffee, and sat in a chair across the aisle from Stretch.

Sound-powered-phone Sailor said, "Fires out. Next launch goes in one hour."

Jon remembered how coldhearted it seemed when it happened this way last year. It had seemed like there was a depraved indifference to the value of a life on an aircraft carrier.

Stretch glanced at Blackey, who radiated cold shoulder, but then he always radiated something unpleasant. LT White had made two combat deployments to Nam and logged more than a hundred missions over North Vietnam. Jon had a hundred missions also, but his were over Laos and South Vietnam. Except for the last mission of last cruise, all his experience was piece-of-cake vice flying into real and lethal danger. The papers called the air defenses over North Vietnam more lethal than that over Germany in WW II. Such hyperbole sold newspapers but ignored history. Odds of

A TICKET TO HELL: ON OTHER MEN'S SINS

WW II bomber crews flying over Germany surviving twenty-five missions equaled a low number. Blackey acted as if he believed the hyperbole, as if the hyperbole defined him, his experience, and his ability. To him, it also defined Stretch's lack of experience and ability.

Stretch wondered how Blackey would behave in the room that night. For sure there'd be some sort of confrontation. *Oh well.* He went back to his letter.

Thirty minutes later, Nose the SDO announced, "Listen up," and he passed out aircraft assignments to the skipper, Nooner, Blackey, and Stretch.

Time to man up. Jon wrote, "Love you," to Teresa and put her, and the letter, in the slide-out drawer in the base of his chair.

He got up, used the head, suited up, checked his aircraft's record, and took the escalator to the flight deck.

The flight deck at night always seemed otherworldly. Muted red light turned it into a Mars scape. But there was nothing muted about the activity going on there. Tractors towed aircraft forward, while others moved aft. If you didn't stay alert and keep your head swiveling, you'd get run over, or cursed for having your head up your butt.

Find your plane, preflight, and strap in, and all the while, Stretch's heart rate eased higher and higher. "Start engines." Check instruments and flight controls for jamming. Check the radar/TIAS scope fastened securely. If it came out of its mount on the cat shot, it would smack into Stretch, and he'd lose control and fly into the water. Scope secure.

On a carrier, death was always close, but it seemed so much closer at night. Pilots said the navy didn't have to pay them anything to fly off a carrier in daytime, but it was impossible for the navy to pay them enough for flying off one at night.

Planes began trundling past, heading for the cats. *Please, God, watch over Teresa.*

At the start of training for a deployment, not only the pilots but the flight deck crew had to learn, or relearn, vital lessons. Operating planes off a cramped, floating airport was a highly perishable skill. Last cruise, and now on this one, the early training carried an air of an eight-year-old boy's first swimming lesson. The lad's father grabs the kid by an arm and the seat of his britches and flings him into a creek. As soon as the youngster surfaces, splashing rather than paddling, coughing, and looking to Pop for salvation, Pop says, "Swim," and turns his back.

A taxi director signaled for Stretch to hold his brakes. It was time to remove the chains securing 511 and him to deck.

"Okay, 511," he mumbled into his oxygen mask, "time to go, and let's see if we can avoid going swimming tonight."

The plane captain hustled across the deck in front of him. Tie-down chains slung over his shoulders and a chock in each hand weighed the skinny kid down.

With a flashlight in each hand, Stretch's director signaled him to come forward, turn left, and switch to a new pair of lights.

So familiar but at the same time so like his first

time at night. Those directors' lights did not seem near enough in the way of orientation. That first time, he'd thought, *What in Sam Hill are you doing here, Zachery? What possessed you to apply for flight training?* It still puzzled him, how he'd gotten through his first night cat shot and night landing.

His director stopped him behind the JBD. The jet blast deflector, a barndoor-sized slab of steel, rose out of the flight deck behind 510 to divert the exhaust from Blackey's aircraft at full power over his own.

Blackey ran up his engine, and the JBD shielded Stretch enough, but it did not prevent 511 from being shaken like a misbehaving child in the hands of an irate parental Godzilla.

The shaking stopped. The JBD dropped, and 510's taillight climbed away. The taxi director motioned Stretch forward and into position. Muted thunks and vibrations through the airframe told of sailors hooking 511 to the cat. His taxi director passed control to the catapult officer. The cat officer turned on his flashlights and signaled, *Feet, off the brakes. Engine, full power.*

Stretch pushed the throttle all the way forward and grabbed the cat grip to prevent the force of the shot from driving the throttle to idle. He stirred the control stick around: no binding. Engine instruments: good. With his pinky finger, he flicked on the light switch on the throttle, signaling *good to go.* He pressed his head back against the headrest. Ice formed on the backside of his belly button. Night cat shots were always high pucker factor—

A silent wham stuffed him hard against the seatback.

He was a bullet zorching down the barrel of a pistol. Then the force let him go. There was the momentary sensation that he would fall, but then the wings grabbed the air. His right hand pulled back ever so lightly on the stick, setting the proper nose attitude, while his left hand let go of the throttle to raise the landing gear. Check to make sure he was climbing. *Good.* Accelerating nicely. Time to raise the flaps.

As he climbed through a thousand feet, 511 accelerated as if propelled by a joy of flying as much as by jet fuel. Stretch felt the fear of dying slither away as he sucked up some of his aircraft's elation. At night, he always felt as if God got him from the flight deck to a thousand feet more than training and experience did.

Climb to overhead beneath a starry sky above a black sea amid firefly lights of twenty other *Solomons*'s aircraft. Manage airspeed and closure rate on the gaggle of lights comprising the skipper's flight. Join on Blackey's wing as number four.

The skipper led them away from the ship for night bombing practice. It wasn't right to say, from that point on, the hop was routine. Night flying never felt routine. It was work. Tension filled, senses on high alert, adrenaline spurting into his blood stream, as if he had a full beer barrel to draw on, until after 511 was chained to the flight deck again. Managing the tension and the adrenaline was, however, familiar. He and 511 were four in the formation, and Blackey wasn't Blackey, he was three.

After the landing, it went in Stretch's logbook as 1.8 hours of flight time. At night. One night cat shot. One night carrier landing. With a purpose code identifying the hop as bombing practice. Ho hum.

Back in the ready room, the skipper debriefed the hop, and the landing signal officer (LSO) gave each pilot his grade. Stretch's landing received a Fair, three points, Blackey an Okay, worth four points.

Blackey stayed in the ready room to write the flight schedule for the next day. Stretch returned to the room with the letter he'd started. As he wrote the first line on the second page, the phone rang.

"Lieutenant Zachery, sir."

"It's Nose, Stretch. Skipper wants to see you in his room."

"You know what for?"

"No. He wants to see you. I know that because he just told me."

Nose hung up.

Now what?

The skipper's room was on the third deck forward, Stretch's third deck amidships, but you couldn't get to the CO's directly. The quickest way was to climb to the hangar, or main, deck, go forward, then descend again to the third.

Knock, knock.

"Come."

The skipper sat at his desk. He motioned for Stretch to sit in the visitor chair. He held out a sheet of paper, a message. "Read this."

It was from the senior chaplain at Lemoore to the commanding officer of VA-92. It concerned Mrs. Zachery.

Jon's heart rate kicked up as high as it had been for the cat shot.

Mrs. Zachery had been hospitalized that afternoon. Nothing was wrong. Mrs. Zachery and her baby were fine, but her doctor wanted her confined where she could be monitored continuously. It was a PRECAUTIONARY MEASURE, only.

How in Sam Hell did "only" fit after two words all in caps?

Jon stared at the base of the door to the CO's room. *Jennifer and EJ!*

"Read the rest of it."

"What?"

"Read the rest of it."

The Warhorse Wives' Group had organized to take care of the Zachery's children. Naomi Engel was taking care of Jennifer and Edgar. It took a second for it to register the chaplain meant EJ.

Plus, Teresa's mother, Mrs. Velmer, would arrive in two days to take care of the children, and the Warhorse wives would continue to provide all necessary assistance. The chaplain would provide daily sitreps.

Jon handed the message back to the CO.

"You know I can't let you go home."

Jon looked at the CO and nodded.

"Okay, Stretch. Meet me in the dirty shirt room for breakfast at 0900."

"Aye, sir." Stretch stood up. "And, sir, could you pass my thanks to Mrs. Fant, to all the wives, and to the chaplain, please?"

Chapter 20

Teresa's room had two beds, the other unoccupied. Most of the time, she looked at company as a blessing. Other times, solitude was nice.

Her first full day in the hospital had seen a steady stream of visitors. The base chaplain brought her communion. All the squadron wives dropped in. Sarah Fant and Laura Davison came together.

Maryann Toliver brought her flowers. Simon Toliver was a slow-talking "aw, shucks, ma'am," guy, while green-eyed brunette Maryann was the obvious alpha in the Toliver pack.

Soon after Maryann departed, Rob White's wife, Carolyn, visited. A very nice lady, sophisticated, cultured. Jon considered her husband to be crude, rude, and abusive. She'd met him at the party. He'd been polite that night. How could Carolyn be attracted to someone Jon called crude and rude?

Teresa's doctor, she called him *the warden*, saw her. "The nurses and hospital corpsmen are here to take care of you. Let them." He wagged his finger at her.

Amy Allison brought her a chocolate shake.

"I'm supposed to watch my weight," Teresa said and sipped on the straw. "I'll watch it tomorrow."

After supper, Teresa called Naomi Engel and spoke with Jennifer and EJ. "Mommy," Jennifer said, "this is just like we talk to Daddy sometimes." But both children were eager to get back to a game they were playing with the Engel children.

Naomi picked up. "Okay, Teresa, we have things under control here. You and your Little Pootzer have a good night."

"Thank you so much for taking them in."

"Oh foot. Nothing to it, and I will tell you, I will miss them when your mother gets here and takes them away from me."

Teresa hung up and wrote to Jon.

At 2200, the nineteen-year-old ward corpsman knocked on the open door. "You need anything, Mrs. Z.?"

She didn't need anything and thanked him for everything he'd done.

"Now if you have to get up in the night, press the buzzer. Doc says we have to help you in and out of bed. Okay?"

"Yes, Mother."

A good kid. Technically not a nurse, but he was a good nurse, and ambulatory, and she confined to bed. Still he conveyed a sense of, "We're in this together, Mrs. Z."

After he left, she finished her letter, picked up her rosary, and turned out her bedside light. She used her "Thank You, God" mysteries.

She thought about Jon on the *Solomons*. Helen, RT's wife, had told her that when a carrier and its airwing started workups for a deployment, both the flight deck and squadron personnel behaved like a T-ball team the first time they ever took the field. "But they get better quickly. At the end of a week," Helen said. "They're pros."

An F-8 pilot crashed the first night. A dramatic example of the distance between T-ball and the pros. She thanked God that Jon flew A-4s and not F-8s. The F-8, and the A-4 for that matter, were old airplanes and would be retired if the Vietnam War would ever end. The F-8 Crusader was harder to land aboard a carrier than the A-4 Skyhawk. Skyhawk pilots called the Crusader "JG killer."

Don't worry, Little Pootzer. Daddy's a good stick.

She smiled at herself. Navy pilot jargon. It seeped into her vocabulary, from Jon but more from the conversation at parties. Her husband didn't talk much about his job and flying, except at parties. She suspected that all the pilots were like that. Only when they were with others who did what they did would the conversation flow. "Beer lubricates rusty vocal cords." Another thing Helen told her.

"Little Pootzer," she whispered. "We have you two days farther along than Daniel. Only twenty-six more to go."

She saw them then, those twenty-six days stacked up against her and each day longer and harder to get through than the last one. Confined to bed. Unable to lift a finger for herself. She even had to summon help to go to the bathroom. So much company throughout the day. At first, she relished the solitude of evening, but that had turned into loneliness of night. Out of the night, guilt swamped her with *You are not where you are supposed to be,* which could be anywhere in the world as long as Jennifer and EJ were in the next bedroom down the hall.

Maryann Toliver popped to mind. *How would she handle being confined to bed for a month?*

And Rose, her best friend since college. Jon described her as four foot ten and ninety-five pounds of don't-take-no-crap-from-*noooobody*. Rose and Teresa's mother were alike in size and demeanor.

For crap sake, Teresa! Rose would say, *If you want to cry, cry already. Then blow your nose, hike up your big-girl panties, and get the crap over it!*

Rose did not take crap from anyone, but she used the word as a way to sneak multiple exclamation points into a single sentence.

Moisture puddled in her eyes. *Rose, it's not crying. My panties are hiked up, and they certainly are for a big girl.* She patted her tummy. *But I am not over it and won't ever be. Little Pootzer and I needed help, and so much of it stepped forward. And we didn't even have to ask.*

Please, Lord, someday give me the opportunity to do something like this for someone else.

Teresa blew her nose. Little Pootzer moved, as if she had just fluffed her pillow and settled into a more comfortable position in bed.

A TICKET TO HELL: ON OTHER MEN'S SINS

Jon left the skipper's room and went to the chapel. He sat on his back-row chair and waited for enough stillness to seep into his soul for him to say, *Thank You, God,* and mean it.

Eventually, he got there and went back to his room. He stopped outside it. Blackey's cassette player blared at top volume. Johnny Cash. "Ring of Fire." Stretch turned the knob, opened the door, and stood there looking at Blackey lying atop his bottom bunk.

"Hey, shithead." He held up a paperback book. "This is a great story. It's about a guy who likes to screw pregnant—"

Jon's ears filled with a swishing sound. He stepped inside and closed the door. He walked over beside Blackey. Blackey's mouth was still moving. Blackey's mouth was always moving.

Jon reached over him, grabbed the battery-powered player, raised it over his head, and smashed Johnny to the deck. The act made no sound that Jon heard. Then he grabbed Blackey's arm, hauled him out of the bed, and punched his roommate in the face. Twice. The second time, Blackey's teeth hurt his knuckles.

Jon released him, and Blackey slumped to the deck, where he sat and leaned against the base of the bunk.

"Hey." Jon nudged him with the toe of his flight boot. "Hey. Clean up the mess. Now. After that, go to sick bay."

The next morning, Stretch placed his aluminum breakfast tray on the table. As he sat, the skipper stared at Stretch's right hand. The knuckles were scraped raw.

"I understand Blackey tripped last night. His shower flip-flops caught on something in the head, and he smashed his face into a sink. Were you there?"

"Yes, sir. I helped him get to sick bay."

Now the skipper studied him with the same intensity he'd aimed at his knuckles.

"So, how're you doing this morning?"

"You're asking if I got my head on straight or am I too worried about Teresa to fly today." Stretch looked at his nasty knuckles, then back up. "When Jennifer was born, I was on a destroyer. Teresa was overdue, and they were going to induce labor on Sunday afternoon. Early that morning, I got a call. I had to get to the hospital right away. Teresa needed an emergency C-section. When I got there, Teresa was on a gurney. She was gray. She was moaning. She'd been in labor for twelve hours and wasn't getting anywhere. Hospitals and weekends, you know. Any rate, that's the *scaredest* I've ever been. I thought she was dying."

"You going to eat that bacon?"

Jon frowned; then he snatched up a piece and stuffed it in his mouth.

"You were saying?"

"So I signed the consent form for the surgery. Teresa was okay. The baby was okay. As soon as I knew that, I went to my ship. We were going to sea the next day, and I asked my boss, the ops O, if I could have leave. He said, 'If I give you leave, what will you do there that your mother-in-law and the other wives can't?' 'I'll be there,' I said. He said, 'What you will be is in the way. No leave.'"

Jon forked up some scrambled eggs, chewed, and swallowed. "In my mind, being there is a thing of great value. But I also believe my ops O was right." Jon held up his left hand and placed the thumb and forefinger a quarter inch apart. "He was right at least that much. And, long way to get to it, but I know the wives are taking better care of things than I could. I'm good to go, Skipper."

Teresa kept her place in her book with a finger and rolled onto her side to check the clock. Sarah Fant and Amy Allison were picking Mrs. Velmer up from the Fresno airport and bringing her directly to the hospital. They should arrive soon.

She anticipated Mother's arrival with relief seasoned with trepidation. The children would have family to look after them. On the other side of the ledger, Father described his spouse as an alpha, with a capital A, female. She never said anything, but Teresa often felt disapproval from her, felt a "You should stand up for yourself more, like I do."

Teresa went back to her book, found her place, and read a sentence; then she looked toward the door.

Mother stood broom handle straight, studying her. Even at thirty years of age, the woman who gave birth to her still made her soul cringe when she did that judgmental scrutiny.

Sarah and Amy were behind her.

"You're okay." It was the kind of sentence she spoke

that might have been an interrogative but sounded declarative.

She marched to the bed. Teresa started rising off her pillows to embrace her, but Mother said, "Stay," and leaned over and kissed her daughter on the forehead.

"I want to see your doctor."

"He comes by in the evening, around seven."

The ladies sat, and a young hospital corpsman stuck his head in the door. "You need anything, Mrs. Z?"

Teresa held up the buzzer. "No thanks, Seaman Evans."

Mrs. Velmer pulled her white gloves off as she watched the young sailor leave. "Cute."

"Mother!"

"Well, he is."

Sarah said, "We set up a duty roster. If your mother needs help, she can call the SDW, the squadron duty wife. I put all our phone numbers on the bulletin board in your quarters. And Amy is right next door. Oh, Naomi has kindergarten duty. She drives her own daughter to and from and will do the same for Jennifer."

"Sarah and Amy have been great," Mother said. "We had a wonderful chat driving from the airport."

Teresa interpreted the expression on her neighbor's face as, *Wonderful may be stretching it.*

Her visitors stayed thirty minutes. Sarah brought Mother back at 1900, just in time to meet Teresa's doctor, which didn't go too badly. Teresa planned to apologize to him in the morning.

Once she'd dealt with the doctor, Mother shifted

A TICKET TO HELL: ON OTHER MEN'S SINS

into her I'm-your-solicitous-parent gear. At 2000, she announced she was ready to go.

Thirty minutes later, Teresa called her neighbor.

"Amy, I am so sorry I put so much on you. And now Mother. She can be a bit much. How are you doing?"

"I'm fine."

"Mother is pushy. So please, take care of yourself."

"I will, and I'm getting along fine with your mother. The one who has trouble with her is Maryann Toliver. She and I brought dinner to your house this evening. And you know how she is. Well, we didn't have all the dinner on the table before your mother put her in her place. And she did it in a way that said, 'I am grateful you brought dinner, but you're in my house now, and here, we do things my way.' I wish I had taken a picture of Maryann's face."

Teresa laughed. "My father says anyone who tries to push Doris Velmer around will get a handful of porcupine quills."

"Everything is fine here, Teresa. Including me. EJ and Jennifer were happy to be in their own beds."

"How will I ever pay you all back?"

"Your mother said that this afternoon. Sarah told her, 'Navy wives don't think of paying back. We think of paying forward.' Good night, Teresa. Sleep well."

After she hung up, Teresa thought about the paying forward notion. There was the saying about sins of the fathers being visited on the sons. Navy wives took blessings and visited them on the daughters. Or something like that. She was too tired to wrestle the mirror image of male sin into female grace.

She thought, *Good night, Jon. Daddy's okay, Little Pootzer. And good night, God.*

Blackey was medically grounded for the first ten days of the at-sea period and stood SDO every day. He'd had an upper incisor bent behind the lower teeth. The dentist straightened the incisor into proper alignment and stabilized the tooth there.

After he got an up chit from the flight surgeon, he flew once or twice a day, just like the other pilots.

In their room at night, Blackey acted as if Stretch wasn't there. He didn't look at him, and he didn't talk to him, which, to Stretch, was the very best way his roommate could behave. Still, Stretch considered him to be stab-you-in-the-back treacherous and was on his toes when they were together. When it was time to sleep, Jon prayed for Teresa and his family and then: *Now I lay me down. Don't let him kill me as I sleep.*

On the last day of the at-sea period, the ship launched an Alpha strike in the morning. Blackey flew as Ironhand two on Stretch's wing. After the flight, CAG called the squadron COs and Stretch to his office for a hot washup.

On the way to the meeting, the skipper asked Stretch how Blackey had done.

"He flew the mission just like I briefed."

"How about designating him as an Ironhand lead?"

"Skipper—"

"Just checking. And you're happy with how you

arranged Ironhanders in front of the Alpha strike, two one mile in front and two, two miles out?"

"That's the way to do it, Skipper. The real proof of the pudding will be in Nam, but as best as I can figure it, this is the way to do it."

"Good. CAG got a message this morning from the bombing range people at Fallon. They are positioning simulated SAM emitters at targets just as you suggested. When we go there next month, they'll be set for us. It'll take the realism in the training a step further. CAG told me to tell you that for a guy with no imagination, a total lack of dedication and ability, you've done a halfway decent job with the Ironhand business."

"You're making me blush, Skipper."

Chapter 21

Teresa's mother visited every other day. Until today, she hadn't been there while Hospitalman Evans was in the room. After the young man left, she said, "Your face is puffy."

Teresa had observed her tracking Evans move around the room like a cat watching a baby mouse. Whenever she went into that hungry predator mode, she always, always wound up saying something embarrassing. She'd expected her to say something about the young sailor.

Puffy face was worse than whatever she might have said about him.

Teresa remembered her high school years as one embarrassment after another. The dominant topic had been her weight. Mother had a way of ambushing her with zingers. "My, doesn't Audrey look nice. No wonder she's homecoming queen." With *no mystery why you aren't* being the real message.

Teresa always believed she'd been fat. Once, Jon had gotten out a picture of her as an adolescent and said, "Not even slightly overweight. See, you look healthy. But! You don't look happy, and that hurts my heart."

Jon also dug out high school yearbooks. In the sophomore edition, he paged to Rosalie Conroy's photo. "Rosalie was fat. See?"

Next, he opened all four yearbooks to Teresa's picture. "See? Not fat!"

Teresa picked up their junior yearbook and paged to Rosalie's picture and pointed to it. Over the summer, their classmate lost adipose tissue from every part of her body except her bust.

"Well, yeah," he said, "there was that."

Teresa swatted him on the arm.

"But I love your bust, Teresa Velmer Zachery. Let me count the ways."

It wasn't only weight. Mother always found something with which to zing her.

When Teresa was pregnant with Jennifer, Mrs. Velmer came to San Diego a week before the due date. The baby was two weeks overdue at her delivery by emergency C-section, which required a month of recovery time. Mrs. Velmer stayed through it all and lobbed zingers about her daughter's clothes, her hair, her makeup.

Jon told Teresa, "She's a good person. Duty called, and she came to help us. I don't think she knows she's being hurtful. Probably considers herself to be frank. And she is driven. Needs to be the center of attention, to be able to say, 'My daughter is homecoming queen. My son-in-law has an important job on his navy ship.' We haven't given that to her, and she can't help being disappointed. But if there's a tough job needs doing, give it to her and get the heck out of the way."

Mother. Salvation and condemnation rolled into one. Often she attached lead weights to Teresa's soul. Her husband levitated it, weights and all.

In a way, dealing with her was like Jon dealing with Blackey. One of the best pilots in the airwing but the most disruptive as well.

Teresa glanced at her mother, who sat mercifully silent. Her knitting needles clickety clickety-ed. One blue bootie complete. Now she worked on a pink one. After the baby was born, she'd make another of the appropriate color. Total labor expended: three booties' worth rather than four. Efficiency. Another of her … whatever.

Mother looked up and elevated one eyebrow above the other. "Jon comes home tomorrow."

Of course, he was coming home tomorrow. Why would she mention that? She recalled how Mother had tracked Seaman Evans. Was she implying Teresa had developed affection for the pimpled, just-out-of-high-school sailor? Dear God. So like her. You expect a particular kind of zinger, and when you think it isn't coming, *zing!*

It will be so good to have him home.

The thought flitted through her head and out again, not even pausing to perch on a branch for a moment.

The timing was rotten. He would be home during the last week of her incarceration. A week later, the squadron would deploy to Fallon, Nevada, for two weeks.

The navy had decided the Warhorses would decommission after the last cruise, then changed its mind.

The squadron had to go back to the war one more time. And not just go back but hurry up to get there. The *Solomons*'s squadrons were given less than half the normal time to train for war. Besides that, the Bureau of Personnel had trouble rousting up enough pilots to fill the suddenly expanded requirements. As if normal times didn't create enough worry over her husband and his duty, there was this frantic urgency over the workups this time.

At least he'd be home when Pootzer was born. If she carried her to term. And not two weeks overdue. Wouldn't that be ironic? Hospitalized for a month to prevent premature delivery, only to have it turn into a late one. If that happened, Jon and the airwing would be back out on the *Solomons*. And this time for a month.

Please, God.

Teresa stopped her prayer of solicitation and instead recited her litany of *Thank Yous*. She thanked Him for her mother. What would she have done if she hadn't been able to come out? It would have put too much on the wives. Even though she was an angel of salvation and the minister of devilish zingers at the same time, she rated a special *Thank You, God*.

Teresa smiled.

The click-click slipped into the background. Thanking Him for blessings evoked contentment, but she'd had to quiet herself to sense it.

After the fly-in from the ship, the Allisons gave Jon a ride home. When he entered his quarters, Jennifer shouted, "Daddy!" and ran to give him an *urka*. He knelt, and over Jennifer's shoulder, he saw EJ standing in the hallway beyond the kitchen, sucking his thumb and watching his father and sister. It had become standard behavior for the boy. Still, each new time, it stuck a needle in his heart.

"Go say hello to your father." Mrs. Velmer nudged EJ's shoulder.

"No!" EJ ran to his bedroom and slammed the door.

Jon stood and walked to Mrs. Velmer. She presented a cheek, and he kissed it.

"EJ always gives me the business after I've been gone. He'll get over it in his own sweet time."

"Do you need something to eat?"

"I'll eat a sandwich. Then I want to get up to the hospital."

Jon entered the kitchen.

"Shoo. Go sit with Jennifer at the table. There's ham. How about a cup of coffee?"

"Yes. Thanks."

"Jennifer is the smartest girl in her class," Mrs. Velmer said from the kitchen. "Her teacher told me as much."

Which meant Mrs. V. put her own interpretation on what the teacher had actually said.

At the table, Jennifer showed her father papers from kindergarten on which she'd written the alphabet, upper and lowercase, and numbers from one to twenty.

After he ate, he went to EJ's bedroom and found the boy sitting on the rug, playing with a truck. "I'm going to the hospital to see your mother. Would you like me to tell her hello for you?"

EJ nodded, and Jon knelt and gave the boy a hug from behind and said, "Urk."

"No." EJ stood and grabbed him around the neck and said, "Urk." When the boy released him, Jon responded, "Ah."

EJ's urk-ah brooked no deviation from his standard. But it seemed as if the boy had consented for his little world to be righted again. Jon kissed him on top of his blond head.

He returned to the kitchen and thanked Mrs. Velmer for lunch and said goodbye to her and Jennifer.

"Dinner's at five thirty," she said.

As he drove to the hospital, he thought about his mother-in-law. She didn't seem so aloof, so cold, so disapproving. Perhaps he hadn't paid close enough attention.

He shoved her off the stage in his mind and anticipated the curtain rising on the star of every drama and comedy that played there. Well, the ones that counted. After parking, he hustled inside and through the hallways to her room.

She was propped against extra pillows, reading. Her hair looked like she'd just come from the beauty parlor. She wore lipstick and a new nightgown. He stood in the doorway and gazed at her.

He couldn't help but compare how she looked now to when he'd found her in the hospital in San Diego the day Jennifer was born. Twelve hours of labor and not making any progress toward delivering their baby had come close to killing her. He'd seen how close death was. She'd been dragged to edge of physical endurance, and she had nothing left to fight with to hang onto her life.

Today, though, she was simply beautiful. The most beautiful pregnant woman, the most beautiful hospital patient, the most beautiful person in the world.

As he stood in the doorway, his eyes drank her in, as thirsty for the sight of her as a wanderer through Death Valley, crawling on his hands and knees, seeing oasis after oasis that turned out to be mirages, finally finding a real one.

She looked up and saw him. The power in her eyes sucked strength and energy from his limbs. "You make my *wees kneak*." And they were weakened.

She opened her arms. "Come here, silly."

He did. He stayed two hours, and they talked about everything, but at the end of the visit, he couldn't remember a single thing they'd said to each other—except "You make my *wees kneak*."

As he pulled out of the hospital parking lot, it was as if he rode a magic carpet, not in a car, but entering Alvarez Village, the carpet lost its magic. He had to face the real Mrs. Velmer.

The one he'd met at lunchtime was not his real mother-in-law. He convinced himself she hadn't had time to administer the proper treatment to her inadequate son-in-law. Maybe she'd even been nice to lull him into a false sense of security. But at dinner, *blamo*, he'd catch both barrels. He saw how it would go. He'd be talking with the children, and something one of them said would trigger one of her zingers. Statements too subtle for the children to understand but aimed at his ... inadequacies, which he would understand perfectly.

After parking in the carport, he shut the engine off. Five twenty-two. There was time.

Mrs. Velmer was, in a way, like Blackey. Impervious to argument. The only thing that got to Blackey was physical punishment when he crossed a line. Jon had never initiated a fight with another person before. In grade school, he'd been in some fights but always begun by the other kid. And he'd, most times, had the snot knocked out of him, until he'd started working on a farm the summer before eighth grade. Then, the farmer's brother, a former US Marine, taught him how to defend himself.

Early in eighth grade, on the way home from school one afternoon, Large Louie, the biggest kid in school and one who got promoted because he outgrew desks rather than because he'd mastered course material, shoved Jon and called him a sissy. Jon spun around and clenched his fists.

Louie grinned. "Sissy wants to fight." He launched a slow, clumsy roundhouse. Jon stepped inside and fired a straight right to Large's nose. The big kid staggered back, looked at the blood on his hands, glared, and charged. Jon ducked the gorilla arms and tripped him. When he rolled over and started getting up, Jon hit him on the ear.

The fight was over, and no one started another with him the rest of the year.

He'd handled Large Louie permanently. Blackey had been, at best, handled temporarily.

Contemplating his roommate resurrected an earlier thought, the one that went, *I never started a fight*. Jon had certainly initiated the physical part of their altercations, but as objectively as he was able, Blackey had been fighting with him, doing his best to smack Jon down with fists of words. He was, he decided, 79 percent confident he wasn't into a game of "You started it," "No, you started it."

Back to Mrs. V. He didn't have a clue as to how to handle her.

Grin and bear it.

Jon got out of the car, walked to the backdoor, and paused with his hand on the doorknob. After sucking in a big breath and huffing it out, he entered his quarters.

Mrs. Velmer stood by the stove, stirring the contents of a skillet. Jennifer and EJ hopped down from the table for *urk-ahs*.

"Wash up. We're ready to eat."

That sounded more like his mother-in-law.

He washed his hands and returned to the table. They said the prayer, and Mrs. V. delivered plates to them and took her seat.

Through the meal and dessert, Jon was alert, watching for the first zinger. One never came. Mrs. Velmer only spoke about the Warhorse wives. She'd gotten to know them all.

Afterward, he ran the children through baths and got them ready for bed. When he returned to the kitchen, Mrs. Velmer was washing dishes. Jon reached for a dishtowel.

"Show me your medal," Mrs. V said.

"Medal?"

"Your medal. Teresa said the navy couldn't decide whether to court-martial you or give you a medal. They tossed a coin, and you lucked out."

He frowned. Teresa didn't speak like that.

"Show me. And bring the write-up thing that goes with it."

He followed orders, brought the medal and citation, and in the conversation that followed, it turned out Maryann Toliver had explained how and why it was awarded.

She read the citation as he dried the dishes and put them away.

Teresa was halfway through the rosary when Jon showed up at eight thirty. He pulled the chair next to her bed, held her hand, and joined her in the prayer.

When they were finished, Jon nodded to the empty bed and told her he was glad she didn't have a roommate. It would be less lonely for her if she did, but when he visited, he liked having her all to himself.

"I haven't had much chance to be lonely. The corpsmen and the nurses have been great. The warden checks on me every day, and every day, someone from the wives' group stops in. Amy and Mother alternate days. I see the kids three times a week. But how are you getting along with her?"

Jon shrugged. "Well. I think. So far, we've had pleasant chats. But I keep waiting for the shoe to drop, you know?"

"I do know. She has never been so pleasant, if I can use your word. It's as if she now respects me. I know she really likes Sarah Fant, and all the Warhorse wives for that matter." Teresa rolled onto her side. "Jon, I am sure Little Pootzer is a girl."

"What? How do you know?"

"I just do."

"Mystical, magical, female intuition?"

"Perhaps."

"There's more?"

"Two days ago, Mother told me we had it tougher than she did during World War II with Daddy in the Pacific. He was gone, letters arrived about once a week. The war seemed so far away. Here, the wives, and especially Sarah, visit Mother a lot. She said one day they talked about the staff car that enters Alvarez Village when a pilot has been killed. Mother said she realized the Vietnam War was right there in Alvarez Village like it never was for her."

Jon placed his hand on the baby. "Ah, Little Pootzerina. You bless our lives even before you formally arrive." LP moved. Jon smiled. "She agrees."

"I'd like to call her Ruth, if you agree."

"Ruth Pootzerina Zachery. Has a ring to it, doesn't it?"

The corpsman, Seaman Evans, appeared in the doorway. "Lights out in five, Mr. Z."

Jon took his leave, but a piece of his heart stayed with

his wife and daughter. As he drove home, he wondered if they shouldn't pick a boy's name too. To be prepared. Or should he try to match Teresa's faith? Then he wondered where he'd sleep that night.

Before he'd left for the hospital, Mrs. V. told him he should sleep in his own bed, that she'd sleep on the sofa. They'd argued the point, and Jon said, "I am not sleeping in the bed with you on the sofa. If you sleep there, I'm sleeping in the car."

When he entered his quiet house, he found a note on the pillow on the sofa. *It's all yours.*

Will wonders never cease?

He'd won an argument with Mother-in-Law Almighty.

Chapter 22

Jon's problem needed resolution before the airwing deployed to Fallon for the next phase of training. It was important that the wing adopt RT's strategy when flying over North Vietnam: leave your onboard jammer in standby until a SAM site locks onto your plane. Blackey called the idea BS. "It means you have to look inside the cockpit when you should have all your attention outside looking for the SAM. And if you're flying Ironhand, you already have to look in the cockpit at that damned TIAS scope."

The problem: the skipper thought Blackey was right.

Try as he might, Jon could imagine only one way to solve it. Probably not the best. Might even be the worst. But it was the only one he could come up with.

He left a note for Mrs. V. and left the house at 0545, arriving at the hospital at six. He spent five minutes with Teresa and then drove to the admiral's office for his meeting with the COS (chief of staff) at 0630.

The chief of staff was a senior captain, the sides of his black hair going gray, flat belly, about five ten. His expression said, *Against my better judgment,* Lieutenant,

I gave you this appointment. It better be worth my while. He didn't invite Jon to sit.

As Jon understood Project Little Round Top, at Lemoore, only Fuller, Morrison, Frost, and he had been briefed into the program, so he had to be careful with how he spoke about it.

"You remember LCDR Morrison, right, sir?" The COS nodded.

"You know he had a space set up to handle a special classified program?" Nod.

"Neither you nor the admiral were briefed into the program, right?"

"Okay, Lieutenant, take a seat. I can see this thing has got you scared shitless. Tell me what you're looking for."

He sat on the edge of the chair. "That program is compartmentalized. LCDR Morrison briefed me into it. My former CO and Frost had been already been read in."

The COS pressed his lips together. "The three stooges." He looked at Jon's uniform shirt. "That commendation medal looks good there."

He didn't want to aggravate the captain, so he thanked him. "Sir, that program has to do with fighting SAMs. Right now, I'm the only one in the squadron briefed into it. It's important stuff, sir. I think my current CO, my current ops O, and another lieutenant, Mike Allison, should be briefed into it as well. And my CAG."

The captain pursed his lips. "Okay, Lieutenant Zachery, I understand what you need." He cracked a tiny smile. "I also understand why it made you piss-your-pants scared. Let me see what I can do with this bucket of shit you dumped in my lap."

A week later, the four Jon requested were briefed into the program. At 1000. The timing stunk. To high heaven. At ten, Teresa was to be released from the hospital. CAG had flown up from San Diego. That set the time for the briefing. Lieutenant Zachery had to attend. He was to deliver half the brief. Second Lieutenant Hu briefed the other half.

Mrs. Fant organized Teresa's jailbreak and busted her out on time.

Jon got home at midafternoon. Teresa and her mother gave Jon a double-barrel EJ treatment. Outside, the temperature was in the high seventies. Inside, frost sat on the afternoon, dinner, and the evening. He deserved the treatment and did his best to absorb it with grace. The children, even EJ, treated him to a warm welcome and all the attention he could handle, which further sank him into guilt and moved frost closer to deep freeze. He tried to get the children to pay attention to their mommy, but they kept coming back to him.

Teresa's special day, and he ruined the crap out of it. He was sure he'd have to sleep in the car that night.

With her toothbrush sticking out of her mouth, Teresa regarded her image. She told herself, *You will not turn out the bedside lamp before you rid your heart of anger. Please, Lord, help me move that mountain.*

When she left the bathroom, Jon was standing there in the hallway.

"Is it okay if I sleep in the bed with you?"

He wasn't her husband. He wasn't a navy carrier pilot who handled night landings and cat shots. He wasn't a man. Only a wounded, contrite little boy. Her heart flushed anger and filled to full with empathy. "In all the world, I wouldn't have you sleep any other place." She patted his cheek. "While you're in there, shave."

"Uh—"

"Shave."

The next morning, Jon closed the doors to the kitchen before he turned the lights on. He didn't want to wake his mother-in-law sleeping on the sofa. But she woke and walked in as he stood by the sink, eating a bowl of cereal.

He'd never seen Mrs. V. this way before. No makeup. A sleep bonnet held on by an elastic band instead of ties. Robe and slippers. And smiling.

"Jon, yesterday, you not being there to take her home hurt Teresa. I understand why you couldn't be there. Sarah Fant explained it to me. I gave you the business because I wanted Teresa to see I was on her side. Understand?"

There he stood with a mouthful of cereal and a grapefruit stuck in his throat. He closed his eyes and muscled the emotional obstacle to swallowing, with its physical manifestation, down. And cleared his throat.

"Are you okay?"

"Yes, ma'am. Thank you for getting up to tell me."

"Just one of the many unappreciated services rendered by mothers-in-law around the world."

She opened her arms. He walked into the hug. Then she let him go. He let her go. She left, returned to the sofa, he presumed. To sleep, he presumed.

The hot-water pot whistled, and he snatched it off the stove, made a cup of instant coffee, peaked in at each of the rooms where sleep was happening, took his coffee, and drove to work.

He was on the flight schedule to lead the skipper, the ops O, and Mike Allison to Nellis AFB, which he did. At Nellis, they were joined by CAG, who'd flown there from San Diego. All were treated to an hour brief summarizing the key points of the Soviet-style air defense system over North Vietnam, with particular attention to SAM operating procedures. Following the brief, CAG and the others were driven to the Little Round Top site. There they observed from inside the site control van as Zachery flew runs against it.

After his runs, Stretch landed and met the others in the Little Round Top building. "Skipper, I briefed you guys that we should go over the beach with our jammers in standby and turn it on only after a site locked onto us. Blackey said I was full of crap. He said we don't want to be fumbling around with switches when we should have our head out of the cockpit looking for MiGs and SAMs. I think you agreed with him. What do you say now?"

"Based on what the intel wienies said, based on what

I saw, I'd say Blackey and I were half-wrong. The point about head out of the cockpit makes sense though."

"What I want … uh, what I *propose* we do, Skipper, is we hold blindfold cockpit drills. Alice and I worked up a scenario, and we tried it on each other."

"It's a good drill, Skipper," Alice chimed in.

"I can show you when we get back," Stretch said. "It'll take ten minutes." He turned to CAG. "Before you fly back to San Diego, sir, can I take five minutes to walk you through the drill? When we're doing Alpha strikes, if the whole airwing doesn't do it the same way, it won't work."

"You want to make it SOP that when we cross the beach into North Vietnam, the whole strike has their jammers in standby mode, and we only turn them on after a Fansong radar lights us up?"

"Right, sir. And if you two heavies agree, I'd like to propose we work jammer switch-ology into the Alpha strikes we fly at Fallon."

"Coupla weeks before Fallon," CAG said, "and you want me to order the airwing to adopt this jammer in standby SOP and to do blindfold cockpit drills before we go to Nevada. That right?"

"Yes, sir."

CAG said, "Lieutenant Zachery, the look on your face says, *The dumb shit heavy finally gets it.* L'il Lord, you might want to spend some time with your boy here and run him through a set of drills that teaches him how to wear a poker face." He rubbed his chin. "Here's what we're going to do. The CO of VF-14, his ops O, and I will fly up to Lemoore tomorrow. Run us all through your cockpit drills. Then, L'il Lord, us dumb shit heavies will talk about it and decide how to handle this." He faced Stretch. "Is that okay with you, Lieutenant?"

Stretch's face got hot. So did his armpits.

CAG laughed. "What'dya think, L'il Lord? Should we change his call sign to Poker Face?"

The next morning, CAG showed up at the Warhorse line with four F-8s. Tiny flew one of them. CAG wanted a JO along so it wouldn't be just heavies imposing a stupid cockpit drill on the airwing.

"Convince Tiny this SOP of yours and the cockpit drill makes sense, and I will make the SOP mine, and I'll mandate blindfold cockpit drills for the airwing."

Stretch briefed Tiny and the fighter squadron heavies for thirty minutes. The briefing consisted of secret material but stayed clear of the highly classified details. After that, he and Alice ran their visitors through blindfold cockpit drills. When those were complete, CAG said, "Tiny, what do you think?"

"Sir, on our last hop, last cruise, we both had SAMs fired at us. I had my jammer on when we crossed the beach into North Vietnam. Stretch had his jammer in standby. I was shot down. He wasn't. My two cents, we should do what these bomber pukes say."

"CAG, in my squadron," Little Lord said, "I have Blackey. He's not going to be convinced. In your other squadrons, there'll be other Blackeys."

"Yep. But as of this moment, jammers in standby is

airwing SOP. It will be briefed, and cockpit drills will be practiced by every pilot in the wing before Fallon. Including your Blackey. This afternoon, you guys brief the other two bomber squadrons up here. Tomorrow, Stretch, you and Alice fly down to San Diego and brief the guys in VF-12. Also tomorrow, work with my ops O and figure out how to incorporate jammer-in-standby drills into our alpha practices for Fallon."

Little Lord assembled the Warhorse pilots in the ready room. CAG, the visiting fighter pilots, plus the COs, the ops Os, and the weapons training officers from each Lemoore-based airwing squadron sat in the back rows of the ready room as Stretch briefed the new airwing SOP on jammers.

When that point was raised, Blackey jumped to his feet. "I've told you this is bullshit. When we go over the beach, we do not want to be fumbling with cockpit switches. We need our heads out of the cockpit, looking for MiGs and SAMs."

"And I've told you, on the Ironhand missions, we will have to manage weapon switches. We'll be carrying both Shrike missiles and bombs. We will have to manage those as the situation dictates. And that means Ironhanders have to have their heads inside the cockpit for a second or two. And the jammer switch is located in a much easier-to-get-to location."

"Well, I ain't doing it that way."

The skipper stood. "Blackey, you will do it exactly as Stretch says, or you won't fly in my squadron. This is airwing SOP."

"That's right, Skipper." Every Warhorse pilot

swiveled around to see CAG. "Jammer in standby going over the beach is SOP. It is vital that everyone accept this and comply with it. Because if one of you," he glared at still-standing Blackey, "decides he isn't buying this, that it's bullshit, you compromise the safety of everyone in the strike. The gomers can tell if you have a jammer on the second they light you up with their radar. If they see a jammer on, they switch to the manual mode, and they can kill you. Much easier than if you abide by our SOP. Jammer in standby is SOP. If any pilot in the airwing says he isn't complying with it, he is off the flight schedule permanently."

"So, Blackey," Little Lord took over, "tell me you you'll comply, or you are, as of this moment, permanent SDO, and you will never fly one of my airplanes again."

Blackey glared at Stretch.

"Look at me," the skipper said. "Say it."

"Aye, sir. I'll follow the rule."

After they said their prayers, after Jon got up from his knees and slipped into bed, Teresa rolled onto her side. "*Whoof.* I am officially changing her name to Big Pootzerina." She took Jon's hand. "Talk to me, Jon Zachery. What's bothering you?"

In the past, *talk to me* was the worst thing to say to him to get him to open up. In bed was the only time they had to talk, and there wasn't a lot of that. Jon was leaving the house at or before six, and sometimes getting home after that in the evening. Tomorrow, he was flying, and he needed to rest, but she could think of no other way to encourage him.

"I am tired, some physically, but my soul is tired. I feel like, since that last hop, last cruise, I've been fighting someone all the time."

Saying something, not a good idea. She kissed his hand.

"That day, when we lost AB and Skunk, I got in a fight with Tuesday. He said something, and I hit him."

She squeezed his hand.

"I was mad at him. I didn't want to stay in the JOB with him there. I slept in my ready room chair for the two nights it took us to get to the PI. Then when everybody was at the O Club, I moved my stuff into Skunk's room. When Dog Lips came back that night, he didn't want me there. He was drunk. I'm sure he was bothered that his plane went down and they launched Skunk instead of him on that last hop." He paused.

She knew what he was doing. He'd dredge up a bit of disclosure and then stop, afraid if he just let loose, he would say too much.

"It wasn't much of a fight. He was too drunk to hurt me, but he went to get the XO, to have him order me out of Skunk's half of the room when he fell down a ladder. The next day, he didn't even remember we'd fought."

Jon took a deep breath and let it out.

"The next two weeks, as we crossed the pond, I fought with myself over what happened to AB and Skunk, what Amos had done to get them killed. Over Tuesday and Dog Lips."

Another breath in and out.

"RT and Tiny and Petty Officer Twombly helped

me inch my way through it. There was so much I didn't want to say, to write to you. It was my burden, and I did *not* want to put any of it onto you."

He rolled onto his side and kissed their baby.

"I felt, I don't know, healed, I guess, by the time we got to Hawaii, but then when I got home, there was Commander Fuller and all that business. I, we, get through that, and now there's Blackey. I want to say, 'Hey, God, enough already.' But of course, I won't say it."

Confession is good for the soul, Jon Zachery.

But she didn't say it.

She did say, "Carolyn White is away for her airline job. She's going to be back in time for the party at the XO's house. Maryann Toliver says Blackey is almost civilized when Carolyn is here."

"If Blackey were married to Maryann, I wonder if she could civilize him."

"Jon Zachery."

"Yes, dear?"

"You talk too much."

Some minutes later, Teresa didn't know how many. To see the clock, she'd have to roll over. Too much work. Her heart thump, thump, thump-ed, like ticktock, ticktock. She listened to her husband sleep, like a baby. Big Pootzerina slept like her daddy did.

Thump, thump. Her eyelids grew heavier and heavier and heavier.

Today went away and took all today's problems with it.

Chapter 23

The TV made noise and kept Teresa company. The children slept. Jon had taken Mother to the party at the XO's house. She'd been tired and thought it best to stay home and rest. He wanted to stay home with her. Mother could ride with the Allisons.

But Teresa knew Sarah Fant intended a surprise for Mrs. Velmer. Probably make her an honorary Warhorse wife. "Take her. I'm okay. Just a little tired."

Jon took her and called after they arrived. The party noise in the background sounded like the canned laughter that burst from the TV now and again.

Teresa sighed. Alone again. Like in the hospital so much of the time. Jon had to go. It was the right thing to do.

Another spurt of canned laughter. *I wish laughter did come in a can.* She pictured opening a can as if it were Campbell's chicken noodle soup and pouring it into—.

Teresa Velmer Zachery!

Friday. Home for five days, and how they'd flown. In the hospital, time dragged so. Monday, the squadron would fly to Fallon and be gone for two weeks. After a brief stay at home, back on the ship. Not long after that, the next deployment would begin. *Maybe there won't be time for Christmas or for you to be born, Big Pootzerina.*

Teresa Velmer Zachery.

She hoisted herself and the baby up from the sofa and headed for the bedroom to get her rosary. In her head, she saw herself waddling down the hallway. As soon as she touched the beads, a calmness displaced the *woe is me.*

She'd written to Jon about how physical contact with her rosary had a spiritual effect on her before she even started the prayers. He responded it was sort of like that with him feeling close to her. Before falling asleep on the ship, he would place his hand against the wall—bulkhead he'd call it—and it made him feel like, through that part of the ship, into the hull, into the ocean, across the ocean, to the California shore, to Lemoore, to her, he could touch her in a spiritual way. But at least partly a physical way too.

He'd written:

> Your rosary beads, Teresa, are like Jacob's ladder. They connect you to heaven and your prayers to God.

He often told her, "You save my soul."
Sometimes, Jon Zachery, you save mine.

Jon figured they'd spend two hours, and he'd still have some of the evening with Teresa. Besides, riding herd

A TICKET TO HELL: ON OTHER MEN'S SINS

on mother-in-law dear for longer than that constituted cruel and unusual punishment.

As soon as they entered the XO's quarters, the skipper's wife took Mrs. Velmer's arm and led her through the crowd to a two-stool bar on the patio. "This is Nat Newsome. The pilots call him Nose. You can call him Bartender. And, Nat, this is Mrs. Velmer, Teresa's mother."

"Pleased. Name your poison."

"A martini with one olive and a raindrop of vermouth, please."

Nose grabbed the gin and prepared the drink.

Mrs. Velmer stirred the olive around, ate it, sipped, and pronounced, "Perfect."

Jon expected to stay latched onto his mother-in-law. To protect her from rowdy pilots packed together where booze was free? Perhaps to protect them from Mrs. V. However, Sarah led her back inside and ushered her around the assembly like a celebrity guest of honor, and she accepted the royal treatment wearing a mien of humble *all this for l'il ole me!* The veneer, however, hid the core of the woman. Royalty suffering the adulation of commoners. *It brightens their day so, you see?*

That was petty, Jon Zachery.

Sometimes guardian angels should neither be seen nor heard.

The skipper invited the patio dwellers inside, then commanded, "Listen up. Sarah has an announcement."

"We've gotten to know Mrs. Velmer over the last months, and we have all been surprised at how readily she fit in. So, I'm making her not an honorary but a full-fledged member of the Warhorse wives. As such, you are entitled to wear this lapel pin."

"Thank you, Sarah. I—"

"Newbies don't get to make a speech," Nose said.

"Bartenders should be seen and not heard," Mrs. Velmer fired back, which elicited some appreciative chuckles and a few catcalls.

"I want to thank all of you for everything you've done to take care of Teresa and the children and for welcoming me. It has been overwhelming. Thank you. Now, how about another martini, Mr. Nose?"

Nose bowed and left. Sarah welcomed Wanda Mason to the Warhorse Wives' Group and presented a pin to her. The skipper thanked Laura for hosting the event. Then he welcomed bachelors Mudder, Cal Mudd, and Stumpy, Walt Short, to the squadron.

"Cal, Walt, welcome aboard," the skipper said.

"Hello—"

Carolyn White clamped her hand over her husband's mouth.

If they'd been in the ready room at the hangar or on the ship, the skipper would have said, "Let's say hello to the new guys." Everyone would respond, "Hello, assholes!"

Laura Davison hosted the party in her house. She was religious. Profanity offended her. The skipper had warned the officers to keep their language clean.

The CO's party dictum overrode the tradition for everyone except Blackey.

"Food's ready," Maryann Toliver announced.

Most of the men moved outside, while the ladies

were served first. The dining room table accommodated eight. Sarah claimed two of the places for herself and Mrs. Velmer and one for Amy Allison. Laura brought plates for them. The Robsey twins occupied the living room sofa with drinks on the coffee table. Apparently, Blackey and Troll had reconciled. The skipper motioned for Jon to join him on the patio.

Outside, Jon found two conversational clumps. Mike Allison and the lieutenants junior grade formed one; the heavies banded in the other. The skipper formed another with Jon.

"Teresa. How is she?"

"She says she's okay, Skipper. Just tired. She insisted I come with her mother. Uh, what your wife did for her. That was really something. I know it means a lot to her."

"Sarah thinks she's one heck of a lady." The skipper shifted gears. "On Monday morning, we fly up to Fallon. You and CAG ops worked out a plan to have us do our first practice Alpha strike on Wednesday. CAG says we launch our first one Monday afternoon. We are jumping right into it. Who do you want as Ironhanders?"

"Me, Alice, Not, and Nooner, Skipper."

"Okay. Now, let's grab some grub before the JO plague of locusts devours it all."

In the kitchen, the skipper filled a plate and returned to the patio. Jon found a chair in the living room opposite the Robsey twins. He wanted to stay close to his mother-in-law at the dining room table behind him.

"Hello, shithead."

Troll said, "Blackey, cool it."

Jon turned. The women babbled happy gabble.

Swiveling back, he ran into Blackey's dark eyes glaring icy hate at him.

It hit him. Blackey wanted the women to hear his dirty language. Especially, he wanted Mrs. Velmer to.

He wants to get at me through her.

He glared back. In the XO's house, he couldn't hit Blackey. The face of the North Vietnamese sailor from 1966—he called him Nguyen—came to mind. Jon didn't hate Nguyen, but he hated Blackey.

Jon placed his plate on the coffee table, went to the kitchen, and used the phone there to call Teresa.

When she answered, Jon said, "You and Big Pootzerina doing okay?"

"Yes. Are you all right? What's wrong?"

"I'm just worried about you two." And he was, but more, he needed to talk to her. Talking to her would lever the hate out of his heart. He told her about her mother getting her Warhorse pin and how she fit in to the Warhorses better than he did.

Teresa said, "When you get home, tell me what's really bothering you. Go back to the party."

He did. When he picked up his plate, he found Blackey still hating him, but at least now hate only lased in one direction, and the enchiladas on his plate looked appealing again.

At a regular interval, male voices and laughter entered the room from the patio, pushing aside, for a moment, the ladies' conversation.

The twins talked about riding dirt bikes. A number of times, Blackey tossed a loud "crotch rocket" into the palaver. Blackey had to push it. Jon turned and found Carolyn White glaring at her husband. He acted oblivious.

The other women didn't notice or ignored it. At the table, Jon observed little dining going on there but much conversation with Mrs. Velmer in the middle of it. He walked out to the patio to get a Jack Daniels on the rocks. The CO, XO, and the rest of the males had circled their folding lawn chairs. They stuffed food in their mouths, chewed once, swallowed, and dove back into the conversational free-for-all.

Jon guzzled a long first sip and then visited the kitchen, loaded his plate with seconds, and returned to the living room. Blackey said to Troll, "You know if Nooner can't go home for lunch, Monica drives out to the hangar and gives him a ..." He leaned over and whispered something to Troll. Two words. The second one was intelligible, "job."

"Knock it off," Troll whispered.

"God's truth," Blackey said. "I saw them. Next time, I'll get a picture."

Lydia Foster jumped to her feet and headed for the front door as if running away from a nightmare. Two of the women hurried after her.

The XO's wife stood in the doorway to the kitchen. Carolyn White said, "I am so sorry, Laura."

Mrs. Velmer marched across the room and slapped Blackey—hard. His plate fell and shattered when it hit the hardwood floor.

The skipper stepped inside. "Blackey. My office. Tomorrow—0600. Leave. Now."

"I'll clean up the mess first."

"I made that mess," Mrs. Velmer said. "I'll clean it up. The one you made can't be cleaned up."

Troll grabbed Blackey's arm and pulled him to his feet.

Mrs. Velmer got on her knees to pick up the pieces. Maryann brought a trash can in from the kitchen.

From the front door, Blackey said, "Come on, Carolyn. We have to leave."

"No. *You* have to leave."

Troll pushed Blackey outside.

Heavy, humid silence filled the room. It was Carolyn's to break. Jon watched her. She shook her head. "I need a place to stay tonight. Can someone put me up?"

"You can stay with us," Amy Allison said.

Jon sat on his easy chair, a plate in one hand, Jack Daniels in the other, and he hadn't done a thing. Inside him, fury burned red and yellow hot from his belly to his brain.

From the mess on the floor, Blackey hadn't eaten much. He'd had a number of drinks. But booze wasn't an excuse. The man was a horse's butt stone-cold sober. On the ship, Jon had promised to hurt him if he crossed the line again. That promise didn't expire when the ship tied itself to a pier.

The celebration part of the party was dead. But all parties, happy or otherwise, left a mess. Carolyn picked up plates to take to the kitchen. Laura Davison reached out to take the load from her. Sarah grabbed Laura's arm and stopped her. Oliver Mason rolled up his sleeves and washed dishes.

Wanda Mason also carried plates to the kitchen and scraped them.

Sarah gathered plastic cups. "Wanda, I'm so sorry this was your introduction to our group."

"Oh, Mrs. Fant. I've seen people drink too much before. And this wasn't the worst I've seen." Jon saw her pause and picture the worst she'd seen. "Rob White sucked the life out of the party, for a moment. But he couldn't take the life from your, *our*, wives' group. The way everyone is pitching in, well, it makes me proud to be a part of it. Besides, I'd have bought a ticket to see Mrs. Velmer in action."

Sarah smiled. "Priceless."

His mother-in-law entered the kitchen and grabbed a dish towel. As she dried, she and Oliver Skippy Mason traded snappy repartee. Soon the washer and dryer had those in the kitchen laughing.

Jon finished his drink as he watched the Masons in the kitchen. Oliver and Wanda went together like a horse and carriage, like their marriage had love in it.

He returned to the patio for his second drink, and as he passed through the living room, he saw Sarah and Amy talking to Carolyn. Sarah held Carolyn's hand. The ladies spoke muted whispers, Sarah's body language eloquent in earnestness and concern.

Sorry, Carolyn.

At the bar, he poured a second Jack.

The skipper stood across the bar from him.

"Stretch, Sarah told me Blackey was looking at Mrs. Velmer when he said what he did. He was trying to get at you through her. But there will be no mishaps in the head. Is that clear?"

"Yes, sir."

Relief flooded through Stretch. Now he didn't have to hurt Blackey.

Teresa was happy to have her husband in bed with her at the end of the day.

Jon had brushed his teeth and gargled. Still, he brought the smell of whiskey with him.

She was happy Mother had had a wonderful time at the party and a spectacular triumph. Sarah Fant had called to check on Teresa and to tell her about the incident with Blackey. Mother was a hero.

Mother was always the hero.

Teresa sighed. She sighed a lot these days. Reaching for her rosary, she brought the beads to their natural resting place, on the baby. A hymn came to mind. It began with a prayer for peace in our hearts, Lord, at the beginning of the day. Stanzas asked for that same blessing at other times of the supplicant's waking hours. It ended with the plea for peace in our hearts, Lord, at the end of the day.

It was there. Peace was. And if you could make your heart be quiet, you could find it.

Chapter 24

Teresa clamped the curling iron onto a tress of her hair. Jon was coming home that afternoon, the trip to Fallon, Nevada, at an end. He'd be home for two weeks, then back out to the *Solomons* for the last bit of training. She wanted to shout, "Enough!" However, the US Navy decided what was enough.

Every day since Sarah Fant broke her out of jail, Teresa had gotten dressed. She would not live in a nightgown and robe all day. Today was special though. Jon was coming home, and she was going to the fly-in.

She put another strand in the curling iron and checked her image. Puffy cheeks.

Mother stops judging me, and now I do it to myself!

Her rosary lay on the counter next to the travel alarm they kept in the bathroom. She reached over and touched the beads and drew comfort from them.

Pootzerina, at thirty-five weeks, was big, but she made her mother huge.

For such a long time, she prayed for the baby to not come early. Now she remembered Jennifer being two weeks late. *Please, Father God in heaven, not that.*

She wanted Ruth to have her birth day. She wanted people to stop fussing over her. She wanted to take care of her children, her house, and her husband.

She put the curling iron aside, picked up the rosary, and held it. The rosary nearby wasn't enough.

In a letter, Jon had written, *Our bodies, our minds, our souls all have muscles, and the muscles of all three have to be kept in shape through exercise.*

The muscles of my body certainly need exercise. In bed in the hospital for a month. *Here at home, Mother won't let me do a thing.*

As she finished her hair, she thought, *Pregnant or not, getting ready in the morning takes time.* Too much time. Time for too much thinking. Too much time for Satan to work his allure on the weakened muscles of her soul.

God won't give us anything we can't handle. She'd always believed that, but now a half-formed thought waited in the wings: *I've handled enough.* But of course, that wasn't hers to decide either.

With her fingers on the first bead, she said the Our Father; then she finished her hair.

The clock said 7:00 a.m. The airwing would fly one last Alpha strike at nine and return to Lemoore at midafternoon.

Even Edgar Jon was excited. It was Mother's doing. She had played up Jon's coming home every day for the two weeks.

I should have done that for our son. But she hadn't seen the way to do it. Perfect Mother saw the way and acted.

It had been a long time since her mother dished

out anything critical, but over those weeks, *I've become witchy.*

She said three Hail Marys, and *Forgive me, Father, for I have sinned. Thy will be done.*

The rosary went into her pocket, and she went to the dining room so Mother could serve her breakfast.

At the table, they held hands and said grace. EJ said his part: "Amen."

Teresa said, "Oh," as she realized something. She was not ready for this baby to be born.

Worry flashed over Mother's face.

"It's all right," Teresa said. "It's just that I have no diapers, no diaper rash ointment, no baby powder."

"You were thinking about Daniel," Mother said. "You had all those things in the house." Mother reached out to take Teresa's hand. "Why don't we ask Amy to pick some up for us?"

"No. I don't want to put any more on Amy. Lydia Foster lives just down the street. I'll ask her to take me. I'm not sure my belly will fit behind the steering wheel."

"Teresa, why don't you tell Lydia what you need? Let her go to the commissary."

"No, Mother. You all have babied me enough."

"Why, Teresa, I didn't know you punned."

The look on Mother's face stunned Teresa.

She's proud of me!

During the first ten days while the airwing was at Fallon, Blackey stood SDO every day. The next day, the skipper put him back on the flight schedule for an Alpha strike. He flew as Stretch's wingman and number two Ironhander. Tomorrow morning, the final day of the Fallon deployment, Stretch wanted him to fly as number two again, only on Alice's wing.

Stretch was in the CO's BOQ room at Fallon.

"You want Blackey to fly on Alice's wing. Does that mean you trust him?"

"Blackey's been on his best behavior since we got up here, Skipper."

"He ought to get divorced more often."

The morning after the party at the XO's house, Carolyn told Amy and Mike Allison that she hadn't seen that viciousness in Blackey/Rob before; she could not live with a man like that, and she was divorcing him.

"This new LT White, he keeps his mouth shut and does his job. But it's way too soon to go all the way to trusting him. Inside his head, there's a lot of thinking going on. If I could see those thoughts, I bet I wouldn't like it. But we're running out of training time, Skipper, and I'd like to see what Alice thinks."

"Okay. And you're flying Wild Card with Tiny."

During the practice Alphas, the Warhorses had been putting up six Ironhand birds. Four flew in front of the formation, and two off to the side. The thinking was the North Vietnamese would focus their attention on the large gaggle of bombers. The two off-to-the side Ironhanders—Stretch called them the Wild Card section—might be mistaken for fighters and sneak up on a SAM site and kill it. Stretch proposed pairing one

TIAS bird with an F-8 for the Wild Card mission. It would conserve TIAS assets.

"Your Wild Card mission seems like the perfect job for Blackey."

"I'd just like to see if this new Blackey act is permanent."

"All right, Stretch. Now get the hell out of here. I'd like to get a little sleep tonight."

"Good night, sir."

The Alpha strike launched into a Fallon-Nevada-clear-to-the-moon sky. Stretch and Tiny took off last.

Above, CDR Fant served as strike lead and orbited over the airfield. Half his twenty planes were still climbing to join the gaggle. To Stretch, the planes against the blue looked like bees swarming around a kicked hive.

The target for the strike was only fifteen miles from the airfield. The plan called for strike lead to fly west for fifty miles, then turn back to the east. Once the gaggle settled onto the heading for the target, the battle exercise commenced, and the gaggle was fair game for the MiG impersonators to attack, and the simulated SAMs would be brought on line at appropriate times by observers on the ground.

Tiny hung on Stretch's wing until the gaggle turned east. Then he assumed the RT formation, above and ahead, and wove back and forth in front of his section lead. The two of them had practiced flying the formation. Tiny said, "This formation has me looking back, not ahead. The hardest thing is trusting a bomber puke—namely you, Stretch—to see MiGs coming at us. I have no doubt you can find SAM sites, but you're a bomber puke, you know. You probably have your eyes locked on the dirt below rather than the sky above, where, you know, MiGs come from."

The trick was *Get Tiny to trust me as much as I trusted RT,* which he admitted to himself would be, well, a stretch.

Tiny flew the formation perfectly, just as he had in their practices. Off to his right, the gaggle was a cloud of gnats.

Forty miles from the target, the early-warning aircraft called, "Bandits over Point Zulu." Zulu was the target. Bandits were MiG impersonators, American A-4 and F-5 aircraft. Adrenaline spurted. It was practice, but the desire to not get shot down was very real.

Good. Tiny stayed in position. Weaving back and forth. The gnats plodded on. And all the while, seconds ticked, and Stretch felt the MiGs, the bad guys, getting closer and closer.

Thirty-five miles from the target. Stretch didn't expect a SAM site to come up until twenty miles to go. Now, thirty miles. The bad guy MiGs. They'd probably go after the gaggle.

"I-five, break right, break right!" Tiny.

I-five was Ironhand five, Stretch. I-five snapped into a ninety-degree bank and yanked back on the stick. He grunted against the g-force, started graying out, and eased on the stick.

Four of them zooming up from down low. Camouflaged A-4s and F-5s, simulated MiGs.

"Fox one on the lead," from Tiny.

"Chaff and flares," from bad guy (BG) one.

Chaff and flares meant BG one had dispensed chaff to spoof radar and a flare to spoof a heat-seeking missile. By the rules, BG one was still alive. Two of the MiGs turned away, heading for Tiny. Two of them headed for Stretch. Stretch pulled down into a dive and passed a BG A-4 and an F-5 climbing straight up.

Stretch rolled right and pulled hard. After reducing his dive angle, he eased off on the stick so he could raise his head and check above and around him for planes. Sky clear. Pulling and grunting, he got wings level. Above and around, not a plane in sight, which meant they were behind him.

"Fox one on the A-4!"

Crap!

"Chaff and flares!" Stretch snapped into a hard left turn. The F-5 pulled up to zoom high. The A-4 rolled to turn with him. Stretch pulled up as if he were going after the climber. As soon as the nose of his aircraft started up, Stretch rolled upside down and pulled hard and headed straight down again. In his head, he had the A-4 behind him, and the F-5 would be back there too but above the A-4. Stretch rolled ninety degrees and pulled hard. Still in a dive, he eased on the stick. Above, clear. Not a plane in sight to the sides. Only one place they could be.

Turn right or left? Right!

He snapped into a hard turn to the right while pulling up. Not a plane in the sky, which meant—

"Guns, guns, guns! On the A-4."

Chaff and flares could not spoof guns. *Should've turned left! If I did, I wouldn't be—*

"A-4, you're dead! Go home."

Stretch pulled to level flight and checked all around for planes. He was game dead, but if he ran into somebody, he'd be the other kind of dead.

To the west, he saw an F-8 behind an F-5.

"Guns on the F-5!" Tiny shot the bad guy F-5.

The A-4 and F-5 who'd killed Stretch were climbing to help their buddies.

The radio was alive with chatter. Other airwing fighters were engaging other MiGs.

Tiny turned to meet the new threat. He passed them going down as they continued to climb. Tiny turned hard, but the two MiGs headed toward the airfield. The remaining F-5 was probably low on gas, Stretch thought.

"Leads in." The skipper, radioing he'd started his bombing run on the target.

Stretch entered the landing pattern at NAS Fallon, thinking there was one advantage to being dead. He'd land first, before the gaggle of pilots did and drained the coffee pot.

CAG held a hot washup with the squadron COs and Stretch. Stretch rode with the skipper to the other hangar.

"I think the bad guys got tipped as to what we were doing with the Wild Card. Four of the MiG simulators pounced on us. They came in low so the early-warning aircraft's radar wouldn't pick them up. Blackey! I bet he did it."

"Nope. CAG tipped them."

"What? Why'd he do that?"

"He wanted *you* to see and appreciate how much you have the Ironhanders hanging it out with your tactics. He likes your approach to the mission but also thought you didn't quite appreciate the MiG risk. The North Viets are no dummies. They will pick up on your approach to Ironhand, and they will make a move to counter it."

"Huh."

"What?" the skipper said. "You don't think heavies can come up with useful stuff like that?"

"Uh …"

The Skipper laughed. "Zachery, you need to spend a half hour in front of a mirror four times a day practicing your poker face. By the way, the fake MiG guys said you were doing a great job holding them off until right at the end. You turned right. If you'd turned left, you'd have defeated their shot, and the F-5 was out of gas. They'd have had to bug out and land. And you'd be alive instead of deader than a doornail." He parked the car and grinned. "Never talked to a dead man before."

As they walked toward the hangar, Stretch replayed the MiG encounter. The first couple of moves he made, he knew where to look for the bad guys when he rolled out of his turn, but the last time, after so many moves, he lost track of where they had to be. It was right or left, fifty-fifty, live or die. He guessed wrong.

Being dead wasn't the worst of it. In the debrief, it came out that Blackey had shot down an A-4 MiG impersonator.

Mother put her hands on her hips. "Sometimes, Teresa Velmer Zachery, you are downright vexatious." She wanted her daughter to stay at home rather than attend the fly-in. "You've done enough today. Shopping. Then getting the cradle in, cleaned up, and set up. You should rest."

The cradle had been in the storage shed in front of the carport. Lydia Foster and Wanda Mason dug it out of the stuffed-to-the-gills *car closet*, as Jennifer called it. Then Teresa and the two young wives cleaned the bed and moved it into the bedroom.

But Teresa would not stay at home. She would go to meet Jon.

Mother and EJ rode to the airfield with Sarah Fant. Her children were in high school.

Teresa and Jennifer rode with Maryann Toliver. They had no children. At the hangar, Maryann parked in the CO's spot, beside which Sarah waited with a wheelchair.

"Sarah, you're too nice to me."

"There's a price. I get to hold the baby right after your mother."

"Deal," Teresa said.

Sarah smiled, and Maryann wheeled the patient through the hangar and to the ramp on the airfield side. Families from the three A-4 squadrons assigned to the airwing lined the length of the massive structure. Maryann set the brake.

The coastal range of hills lining the valley wore its customary baked brown outfit. Those hills were green for maybe two months of the year. Jon would tell visitors, "The San Joaquin Valley is desert. Without irrigation, there'd be no cotton, tomatoes, grapes, what have you."

The hills were brown and dry, but the sun was bright and warm. Early November, pleasant. In a few weeks, the tule fog would set in. Last year, there'd been a solid month with no glimpse of the sun. But that was to face later. Today, the sky was blue. A pleasant breeze blew. And Jon was coming home.

"There," Jennifer said and pointed to the left. Eagle eye, Jon called her.

Teresa heard the jets and saw them as dark clustered spots. Her heart soared. She felt as if she'd been … away, for a very long time. Now, though, she was back where she belonged—at the airfield to welcome her husband home.

The planes drew closer. Jennifer squeezed her hand. "Is that Daddy's plane?"

"No, dear. Daddy's will be one of the last ones to land."

"Do the pilots land in alphabetic order? That's how Teacher calls our names in kindergarten."

Teresa smiled at the earnest face of her bright, inquisitive daughter. "It's something like that."

Chapter 25

At five forty-five, the Monday after the squadron returned from Fallon, Teresa woke for Jon's kiss. He left for work. She began to slip back into sleep with an "Mmmmm" and a smile. He told her he wondered about that moan of ecstasy. Was it because she got to go back to sleep while he had to go to work, or was it because she'd been kissed by a handsome prince? She answered with another moan of ecstasy.

A delicious hour later, she rose, did her morning business, and found Jennifer and EJ at the table. She kissed the tops of their heads and greeted her mother.

Mother paused scrambling eggs at the stove to cheek kiss-kiss her daughter.

Teresa didn't want much to eat and poured a bowl half-full of Corn Flakes. Sliced strawberries on top sounded good. After pouring milk, she took her breakfast to the table. An envelope with her name on it lay next to her plate. Inside:

> You're a very pretty woman
> 'Specially when you're PG.

> It's written on your face and so plain
> to see,
> A mommy's what God made you to be.
> There's just one thing wrong.
> It's gotten kinda hard to hug you
> From frontward on.

She giggled. She chuckled. She laughed—and had a serious contraction. The children were there. She tried not to moan, but a moan escaped, and the wave of pain subsided.

There'd been no buildup. No small ones, so she wondered if it was serious or false labor. No question. Serious. Eye-watering serious. She wiped away the tears with the backs of her hands. Her breath rushed in and out as she tried to get ready for the next one.

"Mommy!" Concern, too much concern, on the face of her little girl.

"Mommy cwying," EJ reported.

Teresa turned to her mother by the stove. In the look they exchanged, Mother understood. She turned off the burners and called Maryann Toliver, the duty wife.

The second one hit. Teresa gripped the end of the table and squeezed her eyes shut. Every muscle in her body tried to rip her apart. It let go. She sagged. Worn out. Already. *Father God in heaven.*

Jennifer frowned. "What's wrong with Mommy?"

Mother smiled. "Your mommy is having a baby."

"Pootz reena," EJ said. "He be a boy."

"I'll pack your things from the bathroom," Mother said.

There was a knock at the back door.

"Jennifer, unlock it please," Mother said.

Jennifer ran, did so, and announced to Amy Allison, "Mommy's having a baby."

EJ uttered his concern. "Breftus."

"Breftus coming right up, Mr. EJ," Amy Allison said.

Maryann stopped her car under the canopied entrance to the base hospital. Seaman Evans waited by the glass doors with a wheelchair. The hospitalman opened the rear door for Teresa.

Teresa looked at Mother. "Did someone call Jon?"

"Mrs. Z." Evans pointed toward the parking lot.

Jon was running toward them.

Evans helped her out of the car and into the chair. Jon knelt on a knee, took her hand, and kissed it. Mother thanked Maryann and closed the passenger door. The car drove away.

Teresa caressed Jon's cheek. "Oh," she said and squeezed her eyes shut, gripping Jon's hand.

"We have to get her inside, Mr. Z.," Evans said, and wheeled her about.

The automatic doors opened, and Evans pushed her into and through the lobby, down a long corridor, and past the wing where she'd been incarcerated.

Hang on, baby. Almost there.

The ride seemed to go on and on. She didn't remember the hospital being so big.

The chair stopped. Evans told Jon that this was an examining room, and he couldn't go in. Teresa squeezed his hand. She wanted him to say, "The heck I can't." But he didn't say anything. He pulled his hand free of her grasp. They wheeled her into the room, and the door closed off her husband.

All I wanted, Jon, was for you to hold my hand. Was that too much to ask?

Two people held her arms and helped her stand. Another took off her slacks and panties. Off came her top. She shivered. "Cold."

They put her into a hospital gown. They lifted her up and onto a table. They placed her feet in cold metal stirrups. "Cold," she said again.

A blanket was placed over her. It didn't help. A ferocious shiver shook her.

Teresa lay there, squinting against the overhead light. Short, shallow, rapid breaths rasped in and out.

"Mrs. Zachery." A nurse stood beside her. "Deep breath in. Let it out. In." Pause. "Out. Good."

Resting, breathing, wondering if time was moving, or had it stopped? She had no idea if the contractions were seconds or minutes apart, but it seemed important to get her breathing done between them.

A nurse appeared, between her legs. Smiled, bent to examine, stood up again, smiled again, and departed.

The contractions continued but grew less powerful, as if her muscles knew Teresa was too small to pass a baby normally, that a C-section was required. But they squeezed moans out of her and dragged her closer and closer to total helplessness. She didn't think she could get up if her life depended on it.

Another nurse arrived and prepped her for surgery. Fourth C-section. Teresa knew the procedures.

A nurse handed her a clipboard with a form, a consent for surgery, with a spinal rather than a general anesthetic. She grabbed the pen and scrawled a signature.

The lights above her sliced a headache across her forehead, which she felt between the contractions. Besides the headaches, there was enduring the contractions and waiting for the next one. And not much else other than thin prayers, which she hoped had more substance and sincerity to them than fear.

They lifted her from the examining room table to a gurney, pushed through swinging doors, down a corridor, and through other doors. Another hoist from the gurney to the OR table.

The place was crowded. Someone hiked her gown up to under her breasts. Another draped a sheet across her thighs. Another painted iodine from her belly button down to the stubble where they'd shaved her.

"I'm Dr. Gibson," said a man through a white mask beneath a white cap and garbed all in white. "You can call me Hoot."

She had nothing to give to that statement.

"Right, then," Hoot said. "On your previous C-sections, you received a general anesthetic. Your OB, your pediatrician, and I agree we should try a spinal. Do you agree?"

"I signed a paper saying I did."

"I'm just trying to be sure."

"I agree. I agree," she said through clenched teeth.

Hands rolled her onto her side. Chill swathed a path from the nape of her neck to her buttocks. Fingers walked down her spine. Counting vertebrae. The fingers stopped. A needle pricked there.

A female voice said, "I'm glad she came in this morning. My daughter has a ballet recital at three."

A male responded, "Ballet? Here in Lemoore, middla-nowhere, California?"

The female said, "Yeah. The teacher and her husband have a small ranch outside town. Her husband raises beefalo, cross between buffalo and beef cattle."

The male said, "Sure hope McDonald's doesn't stick meat like that in a Big Mac."

Hands rolled Teresa onto her back.

Doctor Hoot said something to her. She heard herself respond but had no idea what she'd answered.

Sometime later—it was hard to tell how long because the contractions had stopped—a tiny baby was hoisted into her view. The baby trailed an umbilical cord.

Tiny Pootzerina?

Teresa gasped for air. She thought she was going to throw up. The lights of the world went out.

Jon sat next to Mrs. V. on a Naugahyde-covered bench against the bulkhead in the corridor outside the OR. It passed for a waiting room. The tile glistened with cold white light from the overhead fixtures. The bulkheads were dull green. He rubbed his left hand with the other.

Teresa must have come close to breaking bones with her grip, an indicator of what enduring contractions entailed.

Five years prior, he and his mother-in-law had sat on such a bench waiting for Jennifer to arrive. Mrs. V. took her rosary from her purse, invited him to pray along. She led, and he followed. It felt right to pray like that.

When he waited outside an OR for both EJ and Daniel to be delivered, he waited alone and prayed himself brain-dead. He found himself saying the words, but it was as if he didn't know what they meant. Or maybe he'd squeezed all the meaning out of them.

On the Sunday after they lost Daniel, Jon and the Velmers went to Mass with Teresa in the hospital. The priest spoke about how Jesus prayed during his sermon, "Let this cup pass from Me, but not My will but Thine be done." And on the cross, "Father, forgive them. They know not what they do." Short. To the point. Laden with sincerity and personal investment. Jon agreed with the preacher and decided that was how he'd pray.

As they waited for the arrival of Ruthanne, Mrs. Velmer again invited him to pray the rosary with her.

Father God, who art in heaven, let them live. But not my will, Thine be done.

He responded, "I've done my praying," and opened the book he'd put in the car for this purpose when he'd left for work.

Now there was waiting and Samuel Elliot Morrison's one-volume history of the US Navy in World War II. He read a page, repeated his two-sentence prayer, and got back to the Battle of Midway. Jon thought, as he did every time he read an account of it, *America was lucky.*

True, but they'd also broken the Japanese code. They knew the enemy's intentions.

The double doors to the OR were pushed open. A masked nurse pushed out a cart with a plastic bin on top containing a baby.

Jon and Mrs. V. stood.

"Congratulations, LT Zachery. You have a beautiful daughter."

"Is my wife okay?" Jon asked at the same time Mrs. Velmer asked if the baby was breathing.

"The doctor will be out soon, and yes, the baby is breathing."

It was always the same. Except with Daniel. No nurse had showed up first with a baby. After the waiting, even if you knew the procedure, you wanted to know: How's my wife? But the nurses weren't supposed to comment. You had to wait for the doctor.

"Excuse me, but I have to get your daughter to the nursery."

Jon sat back down as he watched the nurse trundle Ruth away. "They give Teresa an anesthetic for the surgery, and it affects the baby. They talked about giving her a spinal instead of a general. I thought maybe the baby might be awake."

A doctor came out through the double doors. His surgical mask was down around his throat.

"Lieutenant, your wife is fine. We gave her a spinal, but right after we took the baby, she gagged. We couldn't have her throwing up at that point, so we put her under. It'll be a couple of hours before she wakes up and you can see her."

"Doc," Jon said. "Thanks."

Doc shrugged and started walking away, stopped, and turned back. "By the way, Mrs. Allison is in the delivery room. I understand both babies should have a label on them: made in Hong Kong." He grinned.

"You blush so pretty, son-in-law, dear," Mrs. Velmer observed.

"Who's watching Jennifer and EJ?" Jon said.

Maryann Toliver, it turned out, got back to the house just as Amy's contractions started. She called the squadron, loaded the two Zachery children and Amy in the car, and drove back to the hospital.

The doctor permitted Jon to see Teresa at ten thirty. For a half hour. No more. She needed to rest.

Teresa was pale and, as with the other babies, a little detached. She'd explained that a C-section was not only a physical assault she had to overcome but a mental one as well. When she came out from under the anesthetic, she felt like she wasn't all the way back to earth, that wherever she went when she blacked out, she left half herself there, and only the other half was really back.

That explanation made sense to Jon. After they lost Daniel, the same sort of thing happened to him. He would have described it at as being in a mental fog. They lost the baby. The thought did not compute. He could sooner believe the earth was flat. The thought could not get into his head. Some self-defense reflex kept it out. He and Teresa went through the motions, but it was as Teresa described. Both of them were half on earth and half in some other world.

Jon decided neither of them could deal with death when they'd been looking forward to birth. They had, in a sense, shoved the loss into a trunk and stuffed the trunk in the attics of their minds. It would be dealt with later. Just not now. Once he'd reached that conclusion, he congratulated himself. *Way to go, Sigmund. You sure nailed that one.* Whether or not he nailed it, he needed some way to understand what was happening to him.

The most amazing thing, in retrospect, was that the two of them had behaved the same way. The loss hit them as if they'd been crossing a street holding hands, and they were run over by a truck. Instantaneous pain and shock, then numbness. But of course, he hadn't gone through anything like Teresa had. Carrying Daniel for seven months, labor, surgery, a week in the hospital. He had no right to feel the same way she did.

Sometimes, trying to understand a thing to enable you to climb out of a pit of despair, the more you tried, you only wound up digging the pit deeper. And the pit began to feel bottomless, that the pit was worse than despair. It was hell. There was no way out of hell. Except the hell he wound up in after Daniel, at some point he remembered the Our Father, and the second part of it, the ask for things part. He thought he would do like Teresa did and alter the prayer to include a phrase like *pull my soul from hell, God, please.* But he said the prayer and got to *deliver us from evil.* Nothing could be more evil than hell. All the evil from the spiritual world and earthly one wound up there.

Deliver us from evil.

There it was. Just what he needed and already in the prayer. He said it again and felt his personal pit receive a bottom. With something to stand on, he could begin to climb out.

It had taken them a couple of weeks after the loss, but they got to a point where they agreed that Daniel was, and always would be, a part of their family. EJ had a brother, whether he understood it or not.

Now, as Jon drove to the hospital for his half hour, 11:00 a.m. visit, he felt that mental fog roiling inside his head. A part of him was afraid to believe that Ruth was okay, that she would live. *Get a grip, Zachery. You have to be there for Teresa.*

For the entire half hour, Jon held her hand and told her over and over, "You done good, Teresa. Such a beautiful little girl."

"I only saw her briefly," Teresa said, with her eyes open. "They will bring her to me in a few minutes." She smiled at Jon, gripped his hand, said, "Mmmm," and closed her eyes. And melted his heart.

He thought she'd fallen asleep, but her eyes popped open. "Thank You, Father God in heaven, that this isn't like last time."

She closed her eyes, and Jon rapped on his head. No real wood in Teresa's room.

Chapter 26

"Time's up, Mr. Z." Seaman Evans stood in the doorway.

Jon turned. The kid crossed his arms. Doc had said thirty minutes. Evans was giving thirty point zero minutes.

Jon rose and bent and kissed his wife. "I love you." And, "You done good," was thrown in for, well, good measure.

From Teresa's room, he stopped by the nursery and found Mike Allison there gazing through the window.

"Mike," Jon said. "Fancy meeting you here."

"Congrats, Papa," Mike said, and stuck out his hand.

They shook. "You too." They both looked through the waist-high-to-ceiling windows.

There were four infants in the place. A nurse, in an elastic-banded cover for her hair and a mask, tended one at a table at the rear of the space. Two other babies lay in their plastic bins, yowling their tiny heads off and waving their tiny arms. Between them, Ruth slept.

"Remind me to tell Amy to never have quadruplets," Mike said.

"Which—"

"There," Mike cut in, pointing to the one to the left of Ruth. "Amelia. A M from Amy and the M E from Mike."

Jon glanced from Amelia to Ruth, back and forth, back and forth. "Uh, Mike, I gotta go."

Jon returned to the nurses' station in Teresa's wing and asked how to get in touch with Ruth's pediatrician. The nurse gave him the phone number and pointed to the phone atop the counter fencing off her domain.

He dialed the four-digit number. After a female voice answered, he identified himself and asked if he could speak to the doctor.

After waiting, after fingers drumming on the counter, "Doctor Normal."

Jon explained what he'd observed through the window.

"Well, Lieutenant, your daughter, Ruth, is okay. During the C-section, you know they had to administer a general before the umbilical was cut. I am keeping a close eye on her, but we have to give her time for the anesthesia to clear out of her system. Other than being a little sleepy, all her signs are good. Don't worry."

Jon replied, "Thanks, Doc. Sorry to bother you."

He hung up. *Don't worry. Right!*

From the hospital, he drove to the base chapel. He sat in the last pew to the right of the center aisle, closed his eyes, and listened for the silence. Time passed as a continuum thing, not by ticks and tocks, and he heard it. Silence, weighty with solemnity, and dusted with a baby-spoon-sized dollop of peace. In his head, he recited the Our Father.

A noise from the front of church busted apart the silence.

A sailor was setting up the altar for noon Catholic Mass in the all-faith chapel. Twenty before twelve by his watch.

He thanked Father God, who art in heaven, left the chapel, and drove out to the squadron spaces.

After what happened on the last Alpha strike of the Fallon deployment, the skipper said he had been so focused on the anti-SAM mission he overlooked training to counter MiGs. He ordered all the pilots to fly an ACM (air combat maneuver) flight with Blackey. Nose had flown with him yesterday.

Stretch found Nose in the ready room. "How'd your hop with Blackey go?"

Nose shook his head. "Got killed four times. Would've been killed more, but we only had four engagements."

"What did you learn?"

"Not to get in a dogfight with Blackey!"

"What about in the debrief? Didn't he tell you where you made a mistake, where you might have tried another maneuver instead of the one you did?"

"We never debriefed," Nose said.

Next, Stretch talked to Troll. "Did Blackey teach ACM in your training squadron?"

"A little," Troll said. "Only on the couple of hops where a student flew a two-good-guys-fight-two-bad-guys hop. Blackey can beat anybody in ACM. I've never seen or heard of anyone who beat him. But he can't teach for shit. When he flew the two versus two hops, the other bad guy instructor briefed and debriefed the flight."

"Did you teach ACM?"

Troll said he did.

Next, Stretch spoke to Simp, the ops officer, and suggested they needed a new approach to the ACM training the skipper had ordered. Simp listened and took Zachery with him to the CO.

As they walked into the office, the skipper asked, "How're Teresa and Ruth?"

"Both good. Both still sleeping off the anesthetic. Other than that—"

The skipper nodded and shifted gears. "On that last Alpha strike at Fallon, the bad guys diverted the lead Ironhanders from their anti-SAM mission. They had to fight for their lives. Blackey killed one, and Alice fought the other one to a draw." The skipper focused on Stretch. "The bad guys also diverted you from your Wild Card mission. And killed you."

Simp said, "Right, Skipper, but the way Stretch planned the Ironhand for the Alpha strike, he had four anti-SAM birds in front of the gaggle. The lead section got sucked off into a dogfight, but the other guys still did their job, and you got the bombers to the target."

"I'd like to get our missions done without losing anyone."

"Us dead guys buy into that, Skipper."

Simp jumped in. "Having everybody fly with Blackey isn't cutting the mustard. He's a great stick and throttle jockey, but he can't teach. I flew with Blackey this morning. He killed me on each engagement. We got back, and I asked him what I did wrong, what should I have done differently? He couldn't tell me. All he could say is he watched what I did, and after a couple of moves, he knew how to kill me and did.

"Troll used to teach ACM in the training command. I'm going to fly with him tomorrow, see what he's got. Also, Skipper, I spoke with the other A-4 ops officers in the wing and proposed we get F-8 pilots up here to run us through MiG tactics. I'll call CAG ops to set that up."

The skipper nodded. "One other thing. On Thursday, CAG wants a meeting in San Diego for squadron COs, ops Os, and weapons training officers. To plan the Alpha strikes we'll do when we go back to the *Solomons* for the last at-sea period. Our last training opportunity before the shooting starts. He wants to make the most of it."

"Skipper, while I was waiting outside the OR, I reread an account of the Battle of Midway. We won, in large part, because we broke the Japanese code and knew what they were going to do. We won't have that intel on the North Viets. But there is a lot of data on their air-defense system and on the parts of it. Like how they employ SAMs. We need to review everything we have on the air-defense system, the guns, the SAMs, and the MiGs. We need to be prepared for everything they've ever done, and we need to consider what they might try in the future that they haven't tried to date."

"While Teresa was having surgery, you read about the Battle of Midway?"

The CO looked at Simp. Simp shrugged.

The skipper turned back to him. "Today's Tuesday, Stretch. You able to fly down to San Diego on Thursday?"

"I'm good to go, sir. Doc won't promise a hard date, but barring complications, he'll release her next Monday."

The skipper grinned. "She will be released on Monday, and Ruth is to be baptized on Wednesday."

"What!"

"Yep. Sarah called just before you two came in. She, Mrs. Velmer, and Teresa worked it all out. The squadron is invited, by the way. There'll be a reception at the O Club following the church service. And the Allisons' Amelia will be celebrated also. So, we're taking next Wednesday off but working the following Saturday."

The skipper laughed. "By the look on your face, Zachery, I can see you're thinking about trying to regain control of your daughter's baptism. My advice. Stay the hell clear of this thing. Otherwise, you're going to get trampled into the dirt. I'd hate to lose you. Every once in a while, you do something useful around here. Now, you two get the hell outa here. I have to call Mr. Velmer and set up this damned party at the O Club. Sarah told me that's my job. Along with dad-in-law, that is."

"Judas priest!"

Jon was logging a lot of miles on the road to and from the airfield. On his way to the hospital for his afternoon visit, he thought about the baptism. He'd intended to talk with Teresa about having Ruth baptized in the hospital tomorrow.

When they lost Daniel, a priest from the parish in the town of Kingsville, Texas, told them he'd baptized their son before he died. Jon wondered if the hospital gave the priest access to the intensive care facility for that purpose.

Jon couldn't imagine the hospital doing that and spoke with Teresa's doctor. The doc confirmed what he'd thought.

He made an appointment to see the baptizer priest. It turned out the cleric had administered the sacrament through a closed door and hadn't mentioned that little detail.

"Then it wasn't a real baptism."

The baptizer assured him it was very real. It was very much like a deathbed confession. If a sinner sincerely confessed his sins directly to God, even if there was no priest to hear him own his sins and ask for forgiveness, the man would be forgiven.

The baptizer's words, that it had been a *very real* sacrament, did not sit right with Jon. He never told Teresa. If he knew anything about her, she would reject baptism through a closed door, and their unbaptized Daniel would hurt her heart and her soul.

As he drove, he stuck to the speed limit, plus five MPH, but the nondriving part of his brain worried about sleepy Ruth. Vigorous Amelia Allison yowled and waved her arms. She seemed strong and determined to hang onto life. Ruth slept. She wasn't even fighting to hang on to her life.

And he could not allow Teresa to see his concern. Worry over this would not help her recover.

So, Poker Face Jon Zachery, you've got your work cut out for you.

As he drove through the admin area of the base, he prayed for help in keeping worry from Teresa.

Through the doorway to Teresa's room, he saw her crying and stopped as an icicle stabbed his heart.

No! Oh, God, no!

Teresa saw him and held out her arms.

He went to her and embraced her. After a time, he tried to pull away, but she clung to him. So, he stayed, bent over, awkward. What she needed was all that mattered.

When she released him, he dragged a chair next to her, held her hand, and saw her blurry, anguished face.

"I had to hurt her, Jon."

"What?"

"When they brought Ruth to nurse, she was asleep and swaddled in her blanket. To wake her, I had to unwrap her and flick the soles of her feet with my finger. It hurt her! She went from warm and sleep to cold and screaming. *I hurt her.*"

Shit, God!

Not a good idea to get mad at God. Sorry. Or cuss Him. Sorry.

But it was a darned good thing he was already sitting down. *Thank You.*

"Jon," she said. "You thought—"

She touched his cheek. "I'm sorry."

Without words, through their eyes, he stopped being he. She stopped being she. They became they.

They administered healing to each other.

"Hey, neighbors."

Mike and Amy Allison stood in the hall. "Gotta walk," Amy said. "Doctor's orders. How're the Zacherys doing?"

Jon was glad Teresa answered for them.

"And how's Ruthanne?" from Mike.

Jon frowned.

"Come along," Amy said, and pulled Mike's arm.

"Jon, I'm so sorry. I felt Ruth was right but not totally right."

"And Anne is your mother's middle name." He shrugged. "We certainly owe her."

"I wanted to talk to you first. Make sure you were okay with it."

An unkind thought knocked at the door of his mind, wanting in and wanting to be spoken, but he held it at bay. "Ruthanne no-middle-name Zachery."

She wasn't a poker player either. She was relieved; then she levitated one eyebrow. A clear invitation.

"Ruthanne Teresa Zachery. And also RT." He squeezed Teresa's hand. "We are loading our little girl down with a heavy name. Forever she will cart with her the weight of knowing what her mother and grandmother did to bring her into the world. And someday I will tell her what the initials RT mean to me."

Their healing, it turned out, had not gotten away from them.

When he got home, Mrs. Velmer sent Jennifer and EJ to play in their bedroom.

She handed Jon a Jack Daniels on the rocks. "Sit on the sofa." She sat in the easy chair and raised her martini. "Cheers."

They sipped.

The cold fire sliced down his throat and budded into a puddle of warm in his belly.

"I want to explain about the baptism."

"You don't—"

"Tut, tut. I want to explain." She sipped. "I was speaking with Sarah Fant on the phone, and we both realized how big a part the squadron played in bringing Ruthanne"—she paused, giving him an opportunity to intervene—"and you and Teresa to this point. Sarah and I agreed it would be most appropriate to include the squadron in the baptism. And to include the Allisons. Sarah got Maryann Toliver involved. They came over, and we put the baptism and celebration plans together. Then we went to the hospital and spoke with Teresa and Amy. They both loved the idea."

Jon raised his glass to her. "I love it too."

"You're not mad?"

"Would it do any good if I did get mad?"

Mrs. V.'s lips didn't smile much, but her eyes did at times.

"Another drink, Mrs. V.?"

"Yes, please, but call me Mother, or Doris?"

On the way to the kitchen with the empty glasses,

A TICKET TO HELL: ON OTHER MEN'S SINS

he said, "I cannot call you Mother. I will not call you by your first name."

"As you wish, dear."

"Sure is hard to get in the last word around you Velmer women."

"I promise. Once before I leave, I'll let you get the last word in."

He rattled fresh ice cubes into his glass.

She said, "I felt you roll your eyes in there, Jon Zachery."

"Judas priest."

He wasn't sure, but, through the wall, he thought he felt her face smile as she sat in her chair in the living room.

Chapter 27

Ruthanne fell asleep still attached to Teresa's nipple.
The nursery nurse waited on the visitor chair. She rose and took the baby as Teresa buttoned the top of her gown.

"I know you worry about her," the nurse said. "But judging by her diapers, she is getting enough to eat." She laid the baby in the basinet on wheels. Ruthanne did not stir. The nurse looked up. "I bet she is saving her strength for when you take her home. Then she'll cry all night."

Her pediatrician wasn't worried. The nursery nurse wasn't worried. She'd made an effort to lighten the mood. Teresa's mood, however, had a mind of its own about her Little Sleepyhead.

She thanked the woman as she wheeled her tiny daughter away. Everyone—the doctors, the nurses, the hospitalmen—all rendered exceptional care to her and Little Sleepyhead. But the nursery nurse took her lighthearted mood with her.

The bedside phone rang at 6:30 p.m. She answered, and Stumpy, the Warhorse SDO, informed her that the skipper, Simp, and Jon had landed.

A TICKET TO HELL: ON OTHER MEN'S SINS

They'd been to San Diego for a meeting with CAG, to plan the airwing training for the next at-sea period, which loomed in the near future with fresh heartache.

She hadn't laid eyes on Jon yet today. He'd stopped by the hospital at 0545 as she slept. He left her a note:

> Peace in our hearts, Lord,
> At the start of our day.

A line from a hymn they both loved. The stanzas walked through the parts of the day, with a prayer for peace in "our hearts, Lord," at noon, in the afternoon, the evening, and at the end of the day. Jon had seen her that morning, but he hadn't disturbed her *beauty sleep*. However, he had prayed for peace in her heart.

A little of that peace occupied her heart along with a healthy dollop of worry over Little Sleepyhead.

She picked up her book and read a chapter and then felt eyes on her.

"Jon!"

He crossed the distance between them like a ghost in a happy hurry.

They embraced as if they hadn't seen each other for months. They parted and looked at each other, drank each other in.

"You look ... so good," he said. A cloud darkened his sunny smile. "Are you?"

"Of course. I saved my evening walk for you. Shall we go look in on our little girl?"

She moved the covers off her and swung her legs over the side of the bed.

"Do we have to call Seaman Evans to help?"

"Of course not. You're here."

He pulled her slippers out from under the bed and helped her into her robe.

"After each C-section, I am struck by what an ordeal you go through."

She smiled. "I always think I know how it will be, but each time, it surprises me."

She hung onto his arm, more than she did with Seaman Evans or Mother. Down the corridor, step by step, step by careful step. She squeezed his arm to her. "Sorry I'm such a slowpoke."

He slowed down even more. "I like you hanging onto my arm."

"Silly."

Step. Step. Step.

"Jon."

"Uh, oh."

"It's not an uh, oh." *At least I hope it isn't.* Jon hadn't wanted to talk about the baby's godparents before the birth. After, so many things happened so fast. "Seaman Evans told me how to make a long-distance call and have it charged to our home phone. I called Carin Conover."

She felt him flinch. "I'm sorry—"

"No, no. I didn't want to talk about godparents. It … it seemed like I, we, would be taking too much for granted. I was so worried we were going to lose Pootzerina too." Step. Step. "You suggested them as godparents in the letter you wrote telling me you were pregnant. I was happy with the news about our baby and her godparents. But I started worrying."

They turned down the corridor leading to the nursery.

"Oh, me of little faith," Jon said. "I should have trusted in Him like you do. Sorry, God. Sorry, Teresa."

She gripped his arm. "She's awake."

Ruthanne lay there, all bundled up, her arms contained in the blanket. Only her tiny head and her blue eyes moved.

"It's as if she woke for the very first time and is seeing the wonders of the world fresh. Like Adam and Eve on the first day of their honeymoon."

Leave it to Jon to find honeymoon in the Garden of Eden.

Ruthanne's eyes looked right at Teresa, and her milk let down. Babies had a way of lifting your soul to heaven but at the same time connecting your feet in a firm and practical way to earth.

There was a clock on the wall in the nursery that read 8:45 p.m.

"We should get me back."

"So good to find you awake, Ruthanne T. Zachery. Sleep well. See you tomorrow."

Down the corridor leading from the nursery, step, step, step. They turned the corner.

"Stretch would like to point out we're on the home stretch."

Teresa smiled at his effort, if not at the thought itself.

She placed her hand on her tummy. "Stop. I need to rest a minute."

"Are you all right? Should I get a chair out of one of the rooms? Do you need a wheelchair? Should I carry you?"

She shook her head. "You never know when overdoing it is getting close. Let me stand here a minute."

"I'll get on my hands and knees, and you can sit on my back."

"Jon. Don't make me laugh."

"I'm serious."

From the direction of the nurses' station, Seaman Evans hustled toward them with a wheelchair. He stopped the chair, set the brake, got Teresa seated, glared at Jon, and tsk-tsked.

"It wasn't his fault," Teresa said, but all that did was earn her husband another glower from the parental eighteen-year-old.

The hospitalman started the chair moving. "Two walks were enough for today."

Teresa was too tired to respond. Evans was right though. She should have been more careful. After weeks and weeks in bed, everyone taking care of her, she hadn't realized how weak she'd become.

Back at the room, Evans helped his patient back into her bed.

"I'm going to get a nurse to check her incision. You have to wait outside, Mr. Z."

"When the nurse gets here, I'll step outside."

As Jon walked toward his car in the hospital parking lot, he ground his teeth. Reprimanded by an eighteen-year-old seaman rankled him.

What rankled him most was the kid was right. He

should have been there for his wife, not let others worry about her. The vows: in sickness and in health. Just like the oath he took to preserve and defend.

A sailor's priorities: God, country, the US Navy, family, self.

He opened his car door, sat behind the wheel, and looked at the spread fingers of his hand. God, the thumb. USA index finger. Navy, the middle digit. Family, ring finger. Pinky, self. Every time he recalled those priorities he'd learned in boot camp, it hurt to find family and Teresa number four.

He closed his car door, and the overhead light extinguished. He gripped the wheel.

Father God, who art in heaven, I need a little help down here.

Ask, and you shall receive.

The nurse had pronounced Teresa all right and allowed Jon to say good night, but then she shooed him away so the patient could sleep.

Jon backed out of the space and started the car toward the entrance of Alvarez Village.

Ask, and you shall receive.

The Conovers. The night Daniel was born.

Kingsville, Texas. March 1969. Jon and Teresa had gone to bed before the ten o'clock news. Early the next day, Jon was to fly out to the Gulf of Mexico and log six landings on the USS *Lexington*, the US Navy's training carrier, thereby completing advanced flight training. Thereby changing him from the subhuman lifeform status of *student* naval aviator to naval aviator—period. With no stinking, demeaning adjectival demoralizing qualifier in front. Thereby earning him the right to pin gold pilot wings onto his uniform shirt. Not real gold, of course, but in so many ways, so much more precious than gold dug from dirt. He'd earned those wings in the sky.

They'd gone to bed early. They made love, and Teresa slept.

Jon wanted to sleep. He needed to sleep. He needed to be rested, at the top of his game physically, mentally, and morally. Tomorrow, he'd land the F-9 trainer jet aboard a carrier. The F-9 was a Korean War vintage fighter. Heavier, faster approach speed than the T-2 he'd landed aboard the *Lex* at the end of basic training. Much more demanding to land, the F-9 was.

But wasn't that the way of things? You need sleep more than anything else on earth, and is there a drop of it to be found anywhere?

Beside him, Teresa *breeeeeathed* in, *phooooed* out. A lullaby that he was immune to.

An image of the back end of the *Lex* filled his head. He was on glide slope and needed finesse in his left hand on the throttle and in his right on the stick. But he was wound so tight he could feel spastic sparks of electricity coursing through his nerves. His hands jerked and spasmed on the controls. His plane bobbed and wove on the glide slope like a drunken seagull. People ran from the flight deck.

"Finesse, Zachery." His instructor's voice. "Easy on

the controls." The voice so calm, so soothing. "Finesse, Zachery. Finessssss—"

Teresa woke him.

"What? What?" He was in bed. His heart hammered. The *Lex?*

"We have to go to the hospital."

He sat up.

"Jon. We have to go to the hospital."

Teresa leaned toward him. Her bedside lamp was lit.

"Are you sure? You're only seven months."

She grimaced, closed her eyes, and flopped back on her pillow.

He smelled it then. She was sweating. Teresa never sweated. Except when she was in labor.

He threw the covers back, pulled on his jeans, slipped on some shoes, without socks, put on a shirt but didn't button it.

Step one: get her dressed.

From the closet, he took slacks and a top from hangers and laid them on the foot of the bed. She was able to stand. Nightgown: off. Underwear: on. Slacks and top: on. He sat her on the side of the bed. She buttoned her top while he put socks and sneakers on her feet.

Step two: the kids.

Jennifer, three years old. EJ, one. He'd put a jacket on them and would take them with, if he had to, but first, maybe he could find some help.

They didn't know their neighbors. Most of the houses in their neighborhood were rentals and inhabited by college kids.

Jon left Teresa sitting on the bed, rushed outside, stopped in the middle of the street, and looked for a house with the lights on. There, one house across from theirs. Lights on in a front room.

He ran to it and knocked on the door. Fifteen seconds. If no one answered in fifteen, the kids were going to the hospital with him.

At ten potato, a deadbolt unlatched, and the door opened on a woman, about his own age. She wore a robe over a nightgown.

Jon blurted out his predicament. "Could you possibly watch the kids?"

"Of course. Let me get some shoes on and tell my husband. That's your house with the front door open?"

"Yes, and thank you so very much."

He ran across the street and got his car keys from the bedroom. Teresa sat as he'd left her. "A neighbor lady is going to watch the kids. She'll be right here. I'll call the hospital."

"I called them."

"Hello?" From the front door.

Jon hurried to meet his saving angel neighbor.

"I'm Carin, C A R I N, Conover."

A brunette. She looked put together, as if for a high-society slumber party or something.

He introduced himself and led her to the hallway. At the master bedroom, he introduced Teresa. She managed a weak smile and a weak "Thank you."

Jon showed his neighbor the children, both asleep. "When I get Teresa established in the hospital, I'll come back and let you go home."

"Nonsense. You stay with your wife. I'm here as long as you need me."

With Carin on one side and Jon on the other, they got Teresa into the car.

Nurses waited at the ER entrance with a wheelchair. He got three seconds with her before they wheeled her away. After parking, he got directions to the waiting area. And waited. And prayed. And wished he'd brought a book.

At 0330, the surgeon came to the waiting room. Mask down under his chin, hair still covered, no blood on the scrubs. Jon always checked.

"I'm Doctor Martin. Your wife came though the surgery well. It'll be a couple of hours before she wakes up. You can't go in the recovery room. Your son is okay at the moment, but I will be frank. Two months early, it is going to be touch and go with him. He's in what amounts to intensive care. You can't see him either. Why don't you go home and get a shower, then come back."

With Jennifer and EJ, he'd gotten to see them before they were transported to the nursery. There was some closure with that, some tangible sign that the worst was over, that it would only get better from here.

Jon nodded to Stone-face in his clean scrubs. "Thanks, Doctor."

He called Carin and woke her. That, he decided, was better than walking in on her and scaring the crap out of her.

When he got home, Carin had bacon and toast prepared and was scrambling eggs. He hadn't realized he was hungry, but he packed the breakfast away. Then he brushed his teeth, shaved, and got in the shower.

While the water was running, Carin knocked on the bathroom door and hollered that the hospital was on the phone. "They need to talk to you."

With a towel around his waist, he tracked wet footprints into the rug in the bedroom and snatched up the phone from the nightstand.

"Hello."

"Mr. Zachery, your son—"

Daniel had died.

Jon replaced the handset on the cradle. Whatever was on his face, he saw the reaction on Carin's. She wanted to come and hug him. He could tell, but then her eyes scanned him up and down.

He stood there practically naked in his bedroom with practically a total strange woman, and his son had died alone. Maybe Teresa would be awake now, and she was alone to deal with the loss of their baby.

"You need to go back to the hospital," Carin said. "Your wife needs you."

That was Kingsville, Texas. Two years ago.

He drove into Alvarez Village and thanked God for the Conovers.

He'd needed help and hadn't even asked. Just across the street, he found Carin Conover, who was still awake because she liked to watch Johnny Carson.

They were, he knew, the two people on earth meant to be Ruthanne's godparents.

At home, he parked in the carport and took a minute

to bring himself from then in Kingsville, Texas, to now in Lemoore.

Then he went inside, and Mrs. Velmer served him a drink, dinner, and a sermon. He wasn't getting enough sleep, she said. Their training program was important, he explained. It would save lives during the coming cruise.

"So you kill yourself during training instead of during combat. That makes sense to you, does it?"

After the ten o'clock news, Mrs. Velmer went to bed, and Jon made his on the sofa. He settled onto his pillow and prayed, *Peace in our hearts, Lord, at the end of the day.* For Ruthanne, Teresa, Jennifer, EJ, Mrs. V., Mr. V. Teresa's sister, Hope—

Chapter 28

On Friday, Stretch flew ACM (air combat maneuvering) hops with Troll in the morning and the afternoon.

Troll briefed the first flight. "Let's go out there and see what you've got, Stretch. Then we'll come back and debrief and plan the afternoon go. That okay with you?"

"You're flight lead."

They took off and flew to the area north of China Lake and began their maneuvers from neutral setups. Twenty thousand feet, flying parallel courses, a half mile separation. Troll called, "Fight's on," and they turned toward each other, passed head-on, and Stretch entered his planned maneuver.

They had three engagements. After they landed, Troll drew diagrams on the ready room whiteboard showing what he and Stretch had done.

"On the first one, after the head-on pass, you rolled inverted and got your nose pointed straight down. Then you rolled ninety degrees to your left and pulled back to wings level, where you rolled ninety degrees to your left again, and you had me.

"A dogfight is all looping, swirling curves drawn

through the sky. Like a kid's finger painting. Nothing regular about it. You do what you have to do to counter what the other guy does. It was like you didn't care what the other guy, me, did. You broke the sky up into three dimensions, each of the three ninety degrees off from the other two. I tried to stay with you, but I got a little out of sync with you every time you turned ninety degrees. So you got me on the first engagement.

"Did you try to do this with the bad guys who jumped you and Tiny at Fallon?"

"I did."

"But it didn't work with them. Do you know why?"

"Yeah. To fly in those three dimensions, I spend time looking at my instruments, attitude gyro, for instance. You need to go straight up or straight down. When you turn, you turn ninety degrees. During the maneuvers, I keep track, in my head, of what I've done and what my maneuvers should do to the bad guy. I have to do that because I can't see him all that often. At Fallon, I screwed up. On my last ninety-degree turn, I went the wrong way."

Troll rubbed his chin. "If you have time to plan it beforehand, you fly your plan precisely. Like today. But if you're surprised, like at Fallon, you don't have time to think things through, and the last ninety-degree turn you make is a fifty-fifty live or die deal. That about right?"

Stretch gave a half nod, half shrug. An assent.

"You probably have some success with your three-dimensions maneuver the first time you put it on someone. Right?"

A nod.

"Here's my two cents. What you do surprises people. The first time. But if you try the same thing with a guy who's seen your trick, all the guy has to do is to hang back a bit, keep his speed up, and let you do your three turns. Then he pounces. Like I did. And you get killed. Like you did the next two engagements."

"Most people I've flown against didn't pick up on it so quick."

Troll looked like the most sincere garden gnome in the universe.

"Your maneuver is intended to outfox an adversary, to get you into position to kill him. Thing is, over North Vietnam, the likelihood of you killing a MiG is zip-diddly squat. You carry bombs or missiles to kill SAMs. You do not fly a fighter." Troll leaned forward. "Most of the time, you can expect a MiG to make a slashing attack. And maybe he slashes once and moves off to the side and then turns back at you for another slashing attack. If you can make him miss both slashing attacks, you come home alive."

They both ate lunch. Jon called Teresa. Then Troll briefed the afternoon go.

"Two things to remember, Stretch. Speed is life. Losing sight is death."

Troll explained that over North Vietnam, whether you flew as an Ironhand or as a bomber, your plane would be heavy and carry a high drag count. If at all possible, you wanted to hang onto your Shrike missiles and your bombs. Otherwise, the bad guys, the North Viets, won. You could not complete your mission if they made you jettison your weapons to evade their attack. The thing is, if they did indeed use a slashing attack, all you had to do was make them miss on their high-speed attack. If they came back for second slashing attack, the airwing F-8s would jump them, and you would continue on to the target. Another thing, it was easy to lose airspeed and ability to maneuver if you mindlessly pulled on the stick at an encounter with a MiG.

"This is what Blackey does so well," Troll said. "He keeps his airspeed up, he doesn't lose sight of his adversary, and he doesn't spend his plane's energy until it will do some good. He's patient. That's what I want you to do this afternoon, Stretch; be Blackey out there."

Troll laughed. "The look on your face, man. Like the guy in front of you in church farted."

Stretch and Troll flew at twenty thousand feet, north of China Lake, heading east, a half mile separation between them. Brown, barren desert below. Blue sky, milky with thin, high cirrus above.

"Fight's on," Troll called.

Adrenaline spurted, and Stretch's inclination was to pull on the stick as hard as he could, but he managed to tamp down on the raw emotion boiling his blood, and he watched Troll's plane. He could not allow Troll to get an advantage, and he spent energy to keep that from happening. They dove their planes, they climbed, they turned and snapped into rolls, and Stretch didn't beat Troll, but he held him at bay.

On the hop, they had three encounters, and all three wound up in stalemates.

After the last one, Troll led the flight back to Lemoore. Stretch settled into the wingman position and thought about the flight and how his lead had conducted it.

On one hand, flying to stalemates was not very satisfying. You fight to win, don't you? On the other hand, it wasn't that different from what Stretch wanted his Ironhanders to do. It wasn't as important to kill SAM sites as it was to keep them from interfering with the bombers' mission.

His previous flights with Troll, Stretch flew lead and instructed him on the Ironhand mission. It was easy to consider Troll a wingman. After he and Blackey joined the squadron, the natural way to look at Troll was as Blackey's wingman. As a wingman—period. But on this hop, Troll showed he was a better flight leader than Blackey when it came to teaching people lessons that could save their lives in combat.

Jon kissed Teresa good night, left the hospital, and drove home.

He found Mrs. Velmer in the kitchen, chopping salad ingredients. "Your dinner's in the oven. Be just a minute." She raised her cheek for a kiss, which he obliged. He washed up and checked on the children. Sleeping angels.

At the table, at his place, a Jack Daniels on the rocks awaited.

Mrs. Velmer placed a plate in front of him. He said grace silently and "Amen" aloud. She sat at her place, a martini in front of her.

He lifted his drink. "To the best mother-in-law on earth."

He sipped. She didn't. She stared, her eyes reading him.

"What?"

"You probably thought I didn't like you when you started dating Teresa."

He didn't know what to say to that.

"Jon, your mouth is hanging open. Why don't you take the opportunity to put some food in it?"

He closed his mouth.

She said, "Teresa wasn't popular, and she wouldn't put herself forward. I was concerned she was smitten by the handsome boy from St. Ambrose. I worried you were some gigolo who could have any girl he wanted. I needed to be sure you saw how precious my daughter was, that you were committed to her and wouldn't run off if the going got tough and look for easier pickings."

"Handsome?"

"In high school, you made the girls' hearts go pitter pat."

His sweat pores opened. He blushed something fierce.

His supreme discomfit clearly amused her supremely.

"The way I remember high school, whether a boy had his own car was a lot more important than his looks."

"A car was important to some girls. A boy's looks was important to all of them."

She sipped.

He made sure his mouth wasn't hanging open. And waited.

Jon shook his head. Never, in a million years, did he expect to have such a conversation with his mother-in-law.

"Why don't you eat your dinner?"

Meatloaf, mashed potatoes, green beans, his taste buds wept tears of joy.

Over their second drink, they talked about the Zachery children and about Wesley—Teresa's father—and about Teresa's sister, Hope. Wesley was flying out for the baptism. Hope had school she couldn't miss. Safe topics.

Mrs. Velmer went to bed after Jon insisted on washing the dishes. It would cost him four minutes of sleep. He had to do something to contribute to the running of the house. Then he wrote to Teresa. He had to tell her about the conversation he'd had with her mother while it was still fresh in his mind, even though in his second sentence, he wrote he didn't think his mind would let any part of that discussion grow stale.

"Twice now," he wrote, "I've sat with your mother outside an OR as we waited to see what would come out: life or death. And thank You, God. In both cases, only life came out."

He did not mention Daniel, when one life and one death came through the double doors of surgery. Teresa would know he was thinking about their second son.

He closed the letter with:

> You and Tiny Pootzerina Ruthanne come home on Monday. My heart is fixing to bust wide open.

Teresa expected Jon to bring her an outfit to wear for leaving the hospital. Instead, Mother, Sarah Fant, and Maryann Toliver showed up. They entered the room twittering like birds happy to see the snow gone and spring here at last. Seaman Evans was in the room. Sarah shooed him out.

Maryann carried a dress on a hanger.

"That's not my dress," Teresa said.

"Of course it is," Mother said. "We looked in your closet and decided there was nothing appropriate for your homecoming. You only had fitted prepregnancy things and maternity tents."

"The three of us went shopping," Maryann reported.

"Emerald green is a great color on you, Teresa," Sarah said.

"We were going to get you two dresses," Maryann said. "One for homecoming and one for the christening and the party." She looked at Mother. "But your mother said you'd think that was too extravagant."

Teresa watched Maryann and her mother. An interesting dynamic played between the two. Before when they were together, Maryann always seemed to be striving to establish what Jon would call top dog position. Now, though, there was restraint in Maryann's demeanor. Not deference. Yes. Restraint was the word.

"Jon—"

"After you get dressed, we'll call him," Mother said. "Then he'll come and bring you and the baby home."

"This is just like a wedding," Maryann said. "Your bridegroom shouldn't see you in your wedding dress before you are in church." Maryann's face put on its mischievous imp mask. "Except—"

Sarah cut in. "Let's keep our thoughts pure."

"Otherwise," Mother said, "we'll start calling you Blackey Two."

Mother! She really was a Warhorse wife, whether her husband was a member of the squadron or not. And it wasn't the pin Sarah gave her that made her so.

A nurse brought Ruthanne. The baby was awake. Her little lips were puckered, as if for a kiss. Her eyes roamed around, taking in the wonders of the world.

Sarah placed her hand on the bundled infant in the nurse's arms. "For each of my two babies, the first thousand times I saw them after a nap, or for some other reason I wasn't with them continuously, it was as if hands reached inside my chest and squeezed my heart. Each time, it was the most precious sight I ever saw."

"Right," Maryann said, "except when Little Pooperina wakes you up at two in the morning with a dirty diaper and hungry."

Mother said, "That was so touching, Blackey Two."

Blackey Two Maryann rolled her eyes. Then she took the baby, laid her on the bed, unwrapped her hospital blanket, and removed her hospital gown and cap. With practiced, deft hands, she dressed the tiny girl in a pink frothy dress, a pink cap, and the booties

Mother had knitted. Then she swaddled her in a new blanket, a gift from the wives' group. Laura Davison had embroidered the initials of each member along one edge.

"Sarah, it's—"

"Hold it together, Teresa. There's no time for makeup repair."

Jon appeared in the doorway. He wore his funeral/wedding suit. And a tie. He stood there, a statue with eyes, like Ruthann's, drinking in wonders. A wonder. Her.

"You're not in bed."

"The boy's observant," Maryann said.

"Come on, ladies," Sarah said. "Time for us to get out of Dodge."

Jon came to his wife and embraced her.

"I won't break, silly." She squeezed him to her.

"Ahem," from Maryann.

Jon pulled back, and Maryann handed him the baby. Sarah paraded the women out.

"Another gift from your mother and the wives' group," Jon said. "They are allowing us to do this by ourselves, as a family."

"Ahem." Seaman Evans stood in the doorway with a wheelchair.

"Hospital procedure. A big wheels to take you to your car."

"Thank you, Seaman Smedlap."

"Jon! You know his name."

"Sure, he does, Mrs. Z. But you know these hot-stuff naval aviators. Always joking. Seaman no-middle-name Smedlap. It's the navy's generic sailor."

"Well, Seaman Evans, just to set the record straight, my husband is not a hot-stuff naval aviator. It's much more appropriate to call him a red-hot nasal radiator."

Teresa sat in the big wheels, and Evans headed her for the door.

"He does have a significant honker on him, doesn't he, ma'am?"

"The boy's observant," Jon groused.

Chapter 29

Pure elation swept through Teresa as Jon pulled the car away from the entrance to the hospital.

From the moment she knew she was pregnant, it was as if she carried twins: Little Pootzerina and worry she'd lose her. Like she lost Daniel.

The last couple of months of a pregnancy always dragged. This time, being incarcerated and helpless for the seventh month, it was as if the devil slowed time, elongated the worry, and gave Teresa a taste of eternal hell on earth. And even after the delivery, there was worry over Little Sleepyhead.

But that had all passed. Daddy, two hands on the wheel, concentrating on the road, was driving them home. Teresa looked down at her bundle of joy. And if that was trite, phooey, it was truth. Her tiny child looked back at her, and it was as Sarah Fant said. Angel hands entered her chest and squeezed her heart. A lump formed in her throat.

That's where lumps in the throat come from!

Jon placed a hand on her thigh. Through the physical connection, some of the almost unendurable elation drained off to where she could endure it. It was a sacred moment for her and her husband.

From the back seat, EJ said, "I hold *Roof Ann* first."

He'd been so quiet she'd forgotten he was back there. He wanted to hold his sister. That was sweet. Or did he just want to hold her first since Jennifer was at kindergarten? She turned around. The look on his face answered her question.

Jon sat next to EJ on the sofa and placed the baby across his lap. After fifteen seconds of holding time, he'd logged all the baby holding time he could handle. Daddy took her back. Teresa insisted on fixing lunch.

Jon walked around the living room, cooing and swaying side to side, and soon, her little eyelids drooped, opened, shut, and stayed shut. She went down in her cradle without a peep.

After lunch, as Jon and Teresa cleared the table, EJ ran into the kitchen. "Roof Ann wake up."

Jon offered to take care of the baby.

"Roof Ann poopy."

Jon made another offer. "How about if I wash the dishes?"

"I thought you wanted me to take it easy."

"You want to change the dirty diaper, EJ?" Jon suggested.

The boy shook his head and stuck his thumb in his mouth.

Jon stuck his thumb in his mouth.

She laughed. "That won't work either, buster."

Jon's shoulders sagged. He trudged toward the doorway into the hall. He stopped and turned for another communication with his wife: Teresa and the baby were both healthy, and their navy house was their home, and home was a happy place.

Then he went to do his duty—accomplish a three *Judas priest!* diaper change. Despite the Judas priests, changing a dirty diaper was a blessed dash of normalcy for him and his family.

Once the baby was clean, Jon wrapped her in a towel and brought her to Teresa for a bath in the kitchen sink. EJ dragged a chair in from the dining room and climbed up on it to watch.

Jon watched, too, from behind his son.

Teresa supported the floppy little one with one hand and washed her with the other. The washer was confident, gentle, and efficient with the washee.

With Jennifer, Jon had given her a bath—one. After that, he was afraid he'd squeeze her arm too hard trying to hold her upright. Such a loose-jointed, floppy little body all slippery with soap. Fear of hurting her was worse than anticipating a night cat shot.

With Ruthanne, if he had to, he'd do the job, but if Teresa wanted to handle all the kitchen sink baths, he would not argue.

With the baby on the towel, EJ pointed at her belly and wanted to know what *that* was.

"That is a scab from a tiny little baby owee," Teresa said. "It will fall off in a few days, and we'll find her belly button there."

"Oh." EJ hopped down and yawned.

"Ruthanne might like a nap in her bed," Teresa said. "I'd sure like one in mine."

EJ, too, thought a nap was a good idea.

Once all of them were down, Jon folded the baby's bedding from the dryer. "Be prepared to work overtime," he told the appliance.

Jennifer came in the side door. Naomi Engel would have dropped her off after kindergarten. Mrs. Velmer followed Jennifer in. Maryann Toliver would have driven her home.

Mrs. Velmer put her purse on the counter opposite the washer. "You're supposed to fold the baby's things, not wad them. Let me do it."

Jennifer plopped her schoolbag on the floor, rooted around in it, and held up a paper. "Look, Daddy. I got a star."

Jon took the paper. "Well I can see why. All the ABCs and all the one, two, threes are within the lines."

Jennifer held up another paper. "Tomorrow we have to read this story. Will you listen to me?"

"Of course, princess."

Jennifer grabbed Jon by the hand and led him to the sofa in the living room. She sat. He sat. She read, without a mistake and with inflections indicating she understood the words she gave voice to.

"Very good, princess. I bet you get two stars on this."

Mrs. Velmer walked into the dining room from the kitchen. She had a sandwich on a plate and a glass of milk. "Anybody care for a peanut butter sandwich?"

Jennifer bolted off the couch. Tomorrow's reading

assignment fluttered to the floor like a leaf autumn forgot to color.

At the dining room table, Jennifer clambered up onto her chair. *Thank you, Grandma*, poised in Jon's mind, ready to launch, when Jennifer said it without the reminder.

"You're very welcome, dear," from the kitchen.

Jennifer wolfed off a bite, and Mrs. Velmer exited the kitchen with a martini in one hand and Jack Daniels in the other. She handed his to him on the sofa and sat on the easy chair catawampus from him.

"We had a productive working lunch," she said, and described the plan.

Visitors for the christening ceremonies, from both the Allison and Zachery families, would fly into Fresno tomorrow. The ones renting cars and those who needed a pickup were all identified and planned. There was room in the BOQ to accommodate all of them, and each visitor had a room assigned.

On Wednesday, after the bidenominational christening in the chapel, there would be a party at the O Club. Those details, too, had been nailed down. A room was even set aside to change a baby's diaper or do whatever mothers and babies might need a little privacy for. Mr. Velmer and Commander Fant were paying for the party.

Jon had been miffed that the Warhorse wives, including their newest member, Mrs. V., had taken over the christening as something of their own. "It's not even for us and Ruthanne," he'd complained to Teresa. "It's for them."

She'd set it in perspective. "We asked for their help. And they delivered more than we asked for. They adopted us, and we adopted them. They are family."

He thought, *Sometimes you just need to shut up and let people do things for you,* but he wondered what Pop would say to that.

To Mrs. Velmer, he said, "Once again, what would we have done without you?"

He raised his glass. She clinked hers against his.

Teresa stood in the hallway and watched her husband drink with her mother. As if she was one of his pilot buddies.

Jon knew she did not like it when he drank. At a squadron function, at dinner with others who liked wine, those were tolerable. But at home, like this, on this day that started so—

Father God, who art in heaven, why can't there be one day of uninterrupted beauty?

Jagged, fiery thoughts ripped through her head: *I can't trust him. I shouldn't have taken a nap, counted on him to behave himself.* Jennifer and EJ, even Ruthanne, she trusted them to let her know if they needed her. But her own husband, she couldn't trust him to behave himself.

Bringing Ruthanne home, those angel hands had reached inside her chest and caressed her heart. Such a moment of almost unendurable beauty! Now, the devil reached a hand inside her chest and dumped a cup of poison into her heart.

Without making a sound, she returned to the bedroom. Ruthanne still slept. She still breathed. Teresa took her rosary from the nightstand and considered kneeling next to the bed. But the thought of standing up again and what it might do to her incision dissuaded her. She sat and prayed.

Out of the quiet, understanding occurred. She'd found Mother speaking with Jon in a way she'd never spoken with her own daughter. Like adults mutually respecting each other's thoughts. Jealousy. A deadly sin. Sins offended God, added to the suffering Christ endured on the cross.

I am sorry for having offended You.

And selfish. After Mother and the wives' group had done so much for her, after Jon had worked so hard to give her time each day, visiting even when she was asleep and leaving a note, or a letter, or a poem, she'd gotten upset to see him and Mother share a moment, with *What about me?*

Forgive me.

She finished her prayer, checked on Ruthanne again, and slipped out of the bedroom.

EJ jumped up from the rug, administered a hug around Teresa's legs, and returned to his trucks. Jennifer hopped down from her chair and brought her starred paper for Mommy to see.

Jon got up from the sofa and kissed Teresa with "Sorry." She and he knew it was for his whiskey mouth.

Mother was in the kitchen preparing dinner. They embraced, and Mrs. Velmer went back to stirring ground beef into the spaghetti sauce.

At the sight of her four-foot-ten mother standing on her stool at the stove, her hair all done up for lunch with the wives' group, the apron over her Sunday best, and stirring the spaghetti sauce, the angel hands reached into her chest and squeezed, ever so gently but firmly enough to squoosh the lump up into her throat.

"Teresa dear," Mother said, "why don't you get the dishes out of the cabinet and let Jennifer show you how she can set the table?"

Jennifer ran into the kitchen.

"So you know how to set the table?"

"Granny showed me."

"I have her wash her hands first," Mother said.

"Yes, Mother."

That night, in bed, after their prayers, Teresa placed her rosary on the nightstand and took Jon's hand.

"It is so good to have you and RT home. I was afraid to believe it could happen."

"It's going to take some getting used to, our little baby girl being RT."

"Does it bother you, me calling her that?"

"No. It's just as I said. It takes some getting used to. When we first joined the Warhorses, there were RT and AB. I wasn't sure what to make of the call signs business. The training command did not prepare me adequately to deal with fleet squadron aviators. Now, I really have to think that the other RT is, in reality, Robert T. In my head, he's RT."

A pensive quiet settled like a comforter over them in their bed. She waited for what he was framing to tell her.

"When you were giving Ruthanne Teresa—"

"RT."

He squeezed her hand.

"If you insist." He rolled onto his side. "When you were giving RT a bath this afternoon, and EJ noticed the umbilical remnant, I found myself thinking, *I'll believe Ruthanne is really home with us and safe when that umbilical falls off and I can see her belly button.* But after that happens, I know what I would do. I'd put another condition on God, and I'd say, after this new condition is met, then I'll believe. Oh, me of little faith."

"I want to roll onto to my side and hug you, but just the thought of doing so hurts my tummy."

"You stay right where you are. As you may recall, I've had recent practice hugging you from sideways on."

"I would laugh at your funny, but it would hurt."

He hugged her, and she squeezed his arm.

Sighing out a deep breath, he said, "Funny, but of all the ways to find faith, at the end of it all, if I don't contemplate my sins, I haven't really found it."

"His ways are not our ways, Jon Zachery."

"You are such a blessing, Teresa Velmer Zachery. Has my soul told your soul how happy it is to be married to yours, *soulularly?*"

"*Soulularly?*"

"Precisely."

He kissed her. It was a nice kiss, tender, lips touching, melting into each other but mindful of the teeth beneath them. Then she smelled it. He pulled back, smelling it too?

"Judas priest!"

Ruthanne—RT—yowled.

"Judas priest."

Jon threw the covers back and swung his feet to the floor. "She's probably going to need another bath. We don't have enough bedding for this kid."

There was a knock on the door, and Mother poked her head in. "You need some help."

"Thanks, Mrs. V. We got it. Sorry little pooperina disturbed your beauty sleep."

"You're sure?"

"We're sure. Thank you."

The door eased shut and immediately opened again, EJ pushed by Mother. "Roof Ann *cwyin.*"

Jon shook his head, rolled his eyes, and went to work.

At least he didn't say Judas priest, but then he did, and Teresa placed one hand on her tummy and the other over her mouth. It hurt a little to laugh, but there wasn't anything else to do.

Chapter 30

On Tuesday night, Jon flew a bounce hop. Carrier landing practice. Five pilots took off with a light load of gas and, after liftoff, turned directly into the bounce pattern and logged seven or eight touch-and-goes.

Naval air stations positioned a mirror, the same glide slope indicator installed on aircraft carriers, next to the end of their runways. Navy philosophy: each touchdown a carrier pilot made ashore should be practice for one at sea. But prior to moving aboard ship, airwings required a period of intensive practice.

The mirror sprouted a row of green lights to either side. A yellow ball of light, the meatball, or just the ball, showed in the middle of the device. To catch a wire aboard ship and to practice ashore, a pilot flew the ball, worked to keep it dead centered on the green lights. If the ball rose above the lights, the pilot was above glide slope, and unless he corrected, he'd miss all the wires stretched across the flight deck to arrest his landing. If the ball sank below the green lights, the pilot was below glide slope, and his plane would be dangerously close to the steel cliff at the back end of the carrier. During periods of intense training, besides the mirror, an LSO (landing signal officer) stood beside the runway and graded each pass each pilot flew. At the end of the practice, the LSO judged each pilot. Had he mastered the mirror? Would he do a passable job on the boat? If no, the pilot would not be allowed to go to sea with his squadron. Two failed attempts to qualify for landing aboard a carrier might cost a pilot his wings or reassignment to non-carrier-based planes.

In the training command, an LSO said, "Life on the glide slope behind the boat is simple. You have three things to control: keep the meatball centered; keep your plane on the centerline of the landing area; keep your plane on speed. And you only have two things to move to make that happen: the throttle and the control stick. Simple, see?"

Except Jon didn't think it was simple. For carrier landings, practice didn't make you perfect. It made you adequate. Even Blackey wasn't perfect. He scored the highest grades in the squadron, but his grades weren't perfect fives. He knocked down fours regularly with an occasional three and, just as occasional, a five. LSOs awarded Jon threes for the most part, with an occasional four and an occasional two.

That evening as he manned his plane, he said, "Hello, 507. Let's go knock down some threes. And what do you say? Can we squeeze out one four?"

Preflight, climb in, strap in, start up. Ready.

Simp led the other Warhorses to the takeoff end of the runway. Stretch trailed him.

No moon, just stars above Lemoore. The darker

the night, the better the training. Easy training is not good training. Stretch heard that one at some point since he'd become a carrier pilot. And he'd stood on the LSO platform both ashore and at sea any number of times, trying to understand what the LSOs saw in awarding grades. In effect, the pilots who rolled out on the glide slope in perfect position and didn't allow their planes to twitch in roll or pitch or speed or lineup earned the highest grades. That would mean the pilot locked the ball into perfect position on the mirror and held it there for the entire approach.

After takeoff, Stretch followed Simp on the downwind leg and watched him start his descending turn to the glide slope. He began his own descending turn and rolled out a little below glide slope. Corrected and went high. Corrected and went low. Closer to the ground, the air grew bumpy. Just after dark, the earth, still warm from the sun, generated thermals. Normally, keeping the ball centered absorbed most of his effort, but tonight, maintaining lineup and on speed took just as much work.

Five zero seven smashed onto the runway. Stretch added full power and climbed back into the blackness, chasing Simp's taillight. "Judas priest, 507," he mumbled. "We'll be lucky to get a two on that one."

Stretch completed seven touch-and-goes, full stopped, and taxied to the fuel pit for a hot refueling, refueling with the engine running. After taking on gas, he reentered the landing pattern and logged another eight graded passes. In the debrief, the LSO gave him a two on the first pass, threes on the others. During his second bounce period, he and 507 got a four.

Teresa's eyes popped open. A car had entered the cul-de-sac. Breaths of air and anticipation, in and out, in and out. The side door opened.

Jon!

When he flew late, she didn't sink into sleep. She floated on the surface of it. She inhaled deep of air and relief, threw the covers back, and swung her feet over the side of the bed—and winced.

Forgot the incision. She sat still a moment and took stock. *Okay.* Slipping on her slippers and her robe, she walked out to the kitchen.

He stood at the counter pouring a drink. He set the bottle down and kissed her.

"Coffee mouth," he said. "Probably just as bad as whiskey mouth."

She didn't correct him, didn't ask, "Rough night?" She squeezed his arm to her.

He sipped. "A hard day's night."

The working-like-a-dog line came to mind, but she didn't give voice to it either.

Jon looked down at his drink on the counter and spoke to it. "The thing is I am not a newbie. I'm a second cruise guy. I expect myself to be better than the real newbies in the squadron."

Oh. One of the new guys got higher landing grades than he did. She trusted her flash of insight. Words of commiseration, of consolation built in her, pressed her to say them. If she did, however, it would slam a door, and he would never again let his guard down with her.

But what did it mean? Should she worry he didn't get good landing grades? As if there wasn't enough to worry about. *Tiny*. He was an LSO. She'd find an opportunity to talk to him.

"The Conovers and Daddy got here from the airport while you were bouncing," she said. "Also, Tiny arrived from San Diego, and they all got to meet each other before going to the BOQ."

Jon looked at her. His eyes blinked twice. His face smiled. "Teresa Velmer Zachery. Queen of the smooth change of conversational topic. I'd like to kiss you again. If you can stand my whiskey mouth."

She tilted her head back, held her nose with thumb and forefinger, and puckered. His lips touched hers like a butterfly settling on a lily.

They parted. He looked into her eyes, all the way into them. He moved toward her again for another kiss. She put her hand on his chest. "After you brush your teeth."

As he headed for the bathroom, she thought of when he'd been in the first stage of flight training. Most of the student aviators were bachelors, but a handful were married. A chaplain had spoken to the handful of wives and delivered a couple of key messages. Pilots' minds needed to be on flying. Distractions could kill them. So, do not heap worries on them from situations at home. That aspect of his flying did not overly concern her.

In Kingsville, Texas, they'd lost Daniel on Monday.

On Thursday of that week, Jon flew a bounce hop, and the LSO pronounced him good to go to the boat. On Friday, he'd logged his six landings and earned his wings. Afterward, he explained his "hello, airplane" ritual to her, how it shifted his mind into flying only and excluded everything else. He had a way to deal with his nonflying life, but after what he'd said over his whiskey, she worried about something else.

The flight training chaplain had talked about another thing, pilots' competitiveness. Many of the student aviators had been top-notch athletes in college. They were used to being the best. Many aspects of flight training would humble them. The humbling might make them question their ability. The chaplain said, "If your husband says something to that effect, listen to him, but don't say anything unless he asks you to. Confidence in his ability can only come from inside himself."

Did Jon doubt his ability to land aboard a carrier?

Talking to him about the subject was out of the question, but there was Someone Else.

In the bedroom, she took up her rosary, and borrowing a line from the "Navy Hymn," she crafted her "Guard and Guide the Men who Fly" mysteries.

She'd only started around the beads when Jon climbed into bed bedside her.

"Mother and Father, the Conovers, and Tiny are coming for breakfast at seven," she said. "Tiny is grilling pork chops."

He leaned over and kissed her. "Mmmm. Pork chops."

"Mmmm. Colgates," she replied.

Ruthanne fussed.

He sighed theatrically and got back out of bed.

Teresa smiled, as if already her new mysteries of the rosary prayer had been answered. But a cocklebur of worry stuck to the chest of her soul, where the soul's heart would be. She needed to talk to Tiny.

Jon lay on his back, his hand on her hip. She lay on her side and nursed the baby.

He thought about the end of last cruise, after he'd discovered what Tuesday and Botch had done. It was as if there had been a membrane. On one side, darkness, despair, and death abided, and he was stuck there. If only his soul could osmose through the membrane, it would pass into the light, into life, into joy. But he was stuck. RT and Tiny, even Twombly tried to help. Maybe he didn't let them. It took him weeks to work his way through.

Now, though, he wondered if they really did help. It just took a long time. Maybe if he had been on his own with the issue, he wouldn't have beat it.

Teresa whispered, "Jon."

He didn't answer other than by getting up, taking the baby, and burping her. Then he checked her diaper, changed it, and swaddled her up again. She settled into her cradle and emitted a tiny sigh, as if she'd completed a heavy task, as if she'd had a hard day's night.

Day of the baptism. Teresa had asked Jon to wake her at six. Groan. He did. But otherwise, she could not have worked in a shower. And that would not do. So, groan and get to it.

At quarter to seven, she opened the bathroom door and heard Tiny's voice, as well as Jennifer's and EJ's. She found Tiny supervising a working party: Jon and the two children. They all carried armloads of … breakfast?

Tiny. She always thought, *Big*, when seeing him for the first time in a while. Big shoulders and arms. Big hands. Black hair over a narrow forehead and black eyes. Scary, unless you knew to look into those eyes.

If you did, you saw a soul yearning for love, and for something to love. For Tiny, Teresa knew, something to love was much more important than to be loved.

He stepped toward her with his arms open. He stopped and said, "Oh."

She went to him and hugged him, her cheek against his chest. His hands, tentative, restrained, touched her back light as a feather. She stepped back. "So good to see you, Little Man."

He held her shoulders and looked down at her. "Good to see you, Little Woman."

She smiled. The last time they'd seen each other, she'd been eight months pregnant and huge.

Jon and the children paraded in from the patio. Her husband looked at his wife and his friend and kept walking to the carport for another load.

Today would be full, Teresa thought, and tomorrow and Friday just as full. Saturday, everyone would leave. Her opportunity to talk to Tiny was now, and it was ticking away. She followed Tiny through the dining/living room, through the sliding glass door, and onto the small concrete slab patio in the tiny cedar-board-fenced backyard. He ripped open the top of a bag of charcoal, dumped some of it into Jon's grill, and squirted lighter fluid on the pile of briquettes.

"Tiny," Teresa said.

He looked at her, then looked back into the house where Jon and the children were trooping into the living room with another load. Tiny slid the sliding glass door shut. Jon looked at Tiny for a moment; then he herded Jennifer and EJ back into the kitchen.

"That's what I like about your husband," Tiny said. "He can take a hint," and he looked at her, hinting.

"During his bounce hop last night," Teresa said, "Jon didn't get good landing grades."

"He never gets *good* landing grades."

"Last night, one of the newbies got better grades than he did," Teresa said. She explained her understanding of pilot self-confidence and that she worried her husband had lost his.

Tiny scratched a match and dropped it onto the charcoal. It lit with a *phwoof*. He edged the kettle grill farther away from the house. "If I burn down your house today, Ruthanne will never forgive me."

Teresa was miffed. She'd asked a serious question.

The sound of jets approaching the airfield came from the direction of the base admin area.

"You know the nature of our business," Tiny said. "It's not that different from you having babies. If the least little thing goes wrong, you and, or, the baby can die."

She'd wanted him to make her feel better about Jon's grades. Where was this going?

"A good grade means from the moment you roll out behind the ship, the meatball is centered between the green lights. You know what I mean?"

She nodded.

"Jon looks at the mirror, and if the ball is in the center, he says, 'You, communist meatball, I know you are not going to stay there.' It can go up or down; only he can't wait for the ball to decide which way it's going to go. What he does is decide to take control. He adds power and makes the ball go up. He knows he made that happen, and it confirms that he is in charge. Now what he does isn't a big or gross deviation from the normal, but it does cost him a point on his grade. Bottom line for naval aviator Jon Zachery, he is not dangerous, and he reliably lands his airplane in the wires. That's what matters."

"I've heard you and Blackey get good landing grades."

"There are eighty-five pilots in the wing. One of them rolls out in the grove. In five seconds, I know who he is by the way he flies. Jon makes good landings consistently, but he's not going to get good grades because of the way he flies the ball. That may not make sense to you, but it's the way it is."

The sound of rowdy voices came from the kitchen, followed by the Conovers and the Velmers filing into the living room.

She wasn't sure if Tiny had answered her question about Jon's self-confidence. But, just then, a smile was more important. She put one on.

Chapter 31

Teresa and Amy Allison nursed their daughters in the baby room set aside for them at the O Club reception following the christening. Amy sat on a chair. Teresa had occupied a wheelchair since she exited their car in front of the church before the baptisms.

"If I wasn't nursing my baby," Amy said, "I wouldn't get to hold her at all."

"The same with our crew," Teresa replied. "Even the newbie husbands had baby-holding time."

"That's because their wives made them take a turn."

From outside the room, the babble of cheerful voices slipped under the door.

"Such a special day," Amy said. "I wish our little ones could appreciate it."

"Someday they will. Nose was shooting movies from the moment we drove up to the church."

"Movie cameras no bigger than a Kodak Brownie. What'll they think of next?" Amy said.

"It was nice to meet your and Mike's parents. Jon's mother and father don't travel."

"Mike told me about how Jon's father shanghaied him into the navy. Is that story true?"

"Oh, it's true." Teresa put her response on pause to shift sides. "Jon and I started dating the summer before senior year in high school. Most of our teachers were nuns, but a few laymen taught math and science classes. One of them helped Jon apply for an ROTC scholarship. He received a letter saying he did not get an outright appointment, but he was an alternate. The next Saturday, Jon's dad took him down to the navy recruiter office, put a form in front of him, and told him to sign it."

"His dad sounds mean."

"Whenever I visit the Zachery house, he always seems shy and withdrawn, like he wouldn't hurt a fly. When we eat together, he never lets his silverware clink on a plate. It's as if he couldn't stand anyone paying attention to him. But when we brought Jennifer to see them for the first time, I asked him if he'd like to hold her. You know how people say someone's face lit up? That was Jon's father."

"So, he isn't all bad."

"I don't think he's any bad. Jon wrote once that it took him to age twenty-five to figure his dad out, and then only a little. His father, he said, must have had a tough time during the Depression. He never talked about it but seemed obsessed with the idea that Jon and his siblings have it better than he did growing up."

Amy started closing her dress. "A preacher of ours once said, 'There is a little of God's angels in all of us. If we but stop rushing about, and seek, we shall find.' It seems we have to seek hardest with those closest to us."

Teresa smiled agreement.

The door to the baby room opened, and party noise flooded in and ebbed when Teresa's mother closed it.

Mother, at four foot ten, stood between the two taller women, Amy's mother and mother-in-law. All three wore similar suits and pillbox hats.

"Mothers of the bride," Amy said.

"Mothers of the bride?" Mrs. Vclmer said. "Then Teresa's father left the shotgun in the closet way too long."

Amy's mother responded, "But we saved a lot of money on the baptism-wedding two-for-one deal."

"I wonder," Amy said, "if the Warhorse Wives' Group will ever get over the three of you ganging up on them."

"They ganged up on us." Mike's mother pointed to the pin on the lapel of her suit. "They made us honorary Warhorse wives."

Amy's mother pointed to her pin as well.

Teresa glanced at her own mother, expecting to see on her face a smug *They're only honorary; I am a full-fledged member.*

"Of course," Mike's mother said, "we're only honorary, not a regular Warhorse Wife, like Doris. She's been here for two months already and is staying for one more."

Amy's mother took Amelia from her daughter.

Mike's mother rested a hand on Doris Velmer's shoulder. "We've decided to call Doris Big Little Woman." She faced Teresa. "We told Sarah Fant. Sarah said, 'BLW, that's a great call sign for her.'"

"The squadron and BLW," Amy's mother said, "did a wonderful job planning this shindig. There's even a game room for the children."

After the grandmothers burped and changed grandbabies, the seven rejoined the party.

Double D, the XO, stood at a small podium with the party on chairs circled around him. He said, with his parade-ground voice tamped down to a whisper, "They're back. Let's all say hello to the guests of honor."

There was a shuffling and scootching of chairs as everyone raised a glass and whispered, "Hello, Amelia and Ruthanne."

Teresa looked at Amy. A typical group of celebrating aviators would have bellowed their greeting.

"Did you ever expect to see a squadron party like this?" Amy asked.

"Restrained naval aviators. Is that an oxymoron?" Teresa said. "And there's even beer."

"And champagne," Mike's mother added.

"And martinis." *Mother*—excuse me—*LBW*.

Behind the XO, Teresa noticed CAG. Easy to notice him since he was almost as tall as Tiny.

Sarah Fant took CAG by the arm, led him over to the guests of honor and attendants, mothers and grandmothers, and introduced him to them.

Nose, self-appointed head photographer, staged a picture with CAG holding an infant in each arm and flanked by three grandmothers.

The babies tolerated the attention, the passing from hand to hand, the flashbulbs, and the noise until 4:00 p.m. Then, as if coordinated and rehearsed, both began to fuss.

A TICKET TO HELL: ON OTHER MEN'S SINS

The Zacherys and Allisons thanked everyone and took their leave. By the time they got to the parking lot, the noise issuing from the party venue had increased significantly.

Teresa, in her wheelchair, and Amy, walking beside her, looked at each other and smiled.

From the pack of grandparents plus Conovers following them, "Boys will be boys," Mother, LBW, pointed out.

"Sounds to me," Teresa's father said, "like some girls are being girls in there," which drew some chuckles from the Allison family and the Conovers. Mother never laughed at Father's jokes.

They all got to their cars, and Jon helped Teresa inside and handed the baby to her. Carl Mudd, Mudder, took Teresa's wheelchair, folded it, stuck it in the trunk of his car, and hustled back inside.

When they arrived at their cul-de-sac, they found Tiny there. He'd shanghaied trays of food for the Zacherys and the Allisons.

"You all were so busy talking I don't think any of you got a bite to eat," Tiny said.

"And you're probably starving. I bet you haven't had anything to eat for a half hour," Jon said.

Tiny huffed himself up with indignation. "I'll have you know it's been thirty-seven minutes and forty-five seconds. I've wasted away to a mere shadow of my former self."

"If this," Mother gestured, "is a mere shadow of your former self, then, thank You, God, for *tiny* favors."

Father, too, laughed at Mother's funny. And, Teresa

mused, Mother didn't laugh at anyone's jokes. She recalled Jon discovering something about his father, that he'd had it tough during the Depression. Maybe Mother had an experience growing up that made her, well, not afraid, not Mother, but hardened her heart to joy and laughter.

Jon roused her from her reverie and took the baby. "Do you hear what I hear?"

She looked up at him.

"Peace in the parking lot. Do you hear it?"

The Velmers, the Allisons, the Zacherys, the Conovers, and Tiny all were laughing and joking in the cul-de-sac before their quarters.

Five years prior, before Jon applied for flight training, he'd been stationed on a destroyer based in San Diego. They'd gone to church in town one Sunday, and after Mass, as they waited for the crowd to exit the parking lot, Teresa had observed, "Inside, we offer each other the peace of Christ, but it doesn't even last until we get outside."

Their fellow worshipers were ready to run one another over to get their cars on the street and to whatever else they were anxious to get to.

"Yes. Peace in the parking lot. So that's what it sounds like." She stood. "Thank you for pointing it out to me." She placed her hand on Ruthanne. "And thank you and your friend Amelia for bringing it to us."

Mister Velmer helped Jon with the children's baths and getting them to bed. Then he and Teresa's mother left for the BOQ. A half hour later, Teresa went to bed. Even though she'd been in the wheelchair for much of the day, she was still tired. The Conovers left. Tiny stayed another half hour, and he and Jon talked on the patio for another while. Then he, too, left for the BOQ. Jon visited the bathroom, then slipped carefully into bed.

Teresa took his hand.

"I thought you'd be asleep."

"My body is, but my mind isn't," she whispered, then rolled onto her side and pushed her back against him.

He spooned with her.

"Amy and I," she whispered, "told the skipper and Sarah how much we appreciated the party and the squadron celebrating with us. The skipper said he needed to thank us. The entire squadron, not more than five months ago, had been expecting to be doing something else. He said, 'We were all jerked from other places, plopped here, and told to train like crazy.' He said, 'We've been doing just that, going like crazy ever since. It was nice to have an opportunity to see that we're normal people, that we have lives to live.' He took both our hands and thanked us."

Teresa was quiet. He knew she wasn't finished talking, whispering. She was doing the same thing he was, listening for the baby, but Ruthanne was tuckered from all the excitement.

"Even CAG made a point of coming to talk to Amy and me. He thinks highly of Mike."

Jon waited. After a moment, he pushed back.

"You're letting cold come between us," she hissed.

"You let the cold come between us. On purpose," he whispered.

She giggled and whispered, "Such a beautiful day."

He humphed and pulled her close again.

"CAG does think highly of you, too, Jon Zachery. He said your performance is halfway close to satisfactory."

"Then," he whispered, "it truly is a beautiful day. And if we never find peace in a church parking lot, at least we found it once in our cul-de-sac."

"Mmm," she moaned softly as she slipped from his embrace into sleep's.

He liked that sleep held her, renewed her, healed her.

He liked, too, that he and Tiny had gotten a few minutes together on the patio.

Tiny had confessed the real reason for his split fat lip. He'd told everyone that he'd stopped at a Carl's Jr. just north of LA and bought six hamburgers and two large drinks to get him the rest of the way to Lemoore. As he tried to wrestle it all into his car, he started dropping the drinks, and as he bent to save them, he banged his mouth against the edge of his car door. "Didn't spill a drop," he finished his story. Whenever he told it, someone always asked, "Did you hurt the car door?"

What had really happened when he'd stopped for the burgers was he had been accosted by three bikers. Big bruisers, Tiny called them. Leather vests, no shirts, tattoos, leather britches, red bandana hair-rags over greasy locks, and rat's nest face hair. He walked by them sitting on their bikes on the way to his car.

The closest one to his car, a guy with a potato nose,

A TICKET TO HELL: ON OTHER MEN'S SINS

said, "That sticker on your back window says you one of them navy pilot baby killers."

Tiny stopped and made eye contact with each of the three. The two bruisers farther away dropped their gazes, and he placed his food and drink on the top of his car.

A hand grabbed Tiny's shoulder, spun him around, and a fist hit him in the mouth. Tiny lashed back with a shot of his own. Potato nose crashed back into the other two crowded behind him. Before they could get their feet under them again, Tiny clobbered Potato Nose and broke it, the nose. He fell to pavement, blood oozing out between his fingers. The other two were scrambling to their feet when Tiny smacked first one, then the other. And the fight was over.

Tiny grabbed the vest of the second guy he'd hit and said, "Which way you shitheads going?"

"North."

"Was I you," Tiny said, "I'd go south."

Tiny let go of the guy and put his food in his car and got in.

Jon, still spooned, took in a deep breath and let it out, being careful so the exhale wouldn't tickle her neck. The antiwar protest. There might be a good side to it somewhere. But when he bumped against it, or it came close to people he loved, he sure could find no good side to it.

On the other hand, the Conovers. In Kingsville, he desperately needed help. He walked into the street in the middle of the night looking for a light on some place and found one. He knocked on the door, and Carin Conover said, "Of course I'll help. I'll be right over."

Be nice if the Conovers defined America.

Of course, they defined a part of it. The quiet part. Meanwhile, the protest part of the country was shouted from the TV and blared in big headlines in the papers.

Another breath in. A breath out.

Thank You, Father God in heaven. I have my family, good friends, and I serve in a good squadron.

With Blackey.

The CO had assigned him as squadron duty officer during the baptism.

Would have been nice to get through a day without thinking about him.

Sorry, God.

Chapter 32

Thursday morning, 0800, Stretch, with Alice right behind him, rapped on the skipper's office door.

"Come."

He opened the door to find CAG, the XO, and Simp, the ops O, all seated and squeezed into the room with a vacant chair for the two of them.

"Sit."

They sat.

CAG said, "Ever since you got shot down in Fallon, your skipper and I have been talking about the Ironhand mission."

The way Stretch had set it up made sense, CAG explained, right up until the shoot down. Nobody knew as much about SAMs and TIAS or had as good a feel for the mission as Stretch and Alice. Most of the time, they would fly Ironhand lead, and a handful of handpicked junior officers would serve as their wingmen. "On that Alpha strike in Fallon, you flew Wild Card, and Alice flew Ironhand lead. Hell, the bad guys could have shot you both down, and we'd have been out of Ironhand Schlitz."

"It's time to modify your approach, Stretch," the skipper cut in. "We want you to train every Warhorse pilot, from Simp on down, to do the mission. The XO and I"—the skipper looked at CAG—"need to lead strikes. Otherwise, I'd have you train us too. We want you to train everybody from Alice and those senior to him to lead Ironhand missions, and to let them lead one of our Alpha strikes during the upcoming at-sea period."

From CAG, "We'll treat Ironhand lead like we do division and section lead. You'll designate pilots to be Ironhand leaders once you're convinced they can hack the program."

The skipper again. "You and Alice talk about this, work up what needs to be done, and get back with CAG and me after lunch."

"Judas priest!" Stretch shook his head. "Friday, Saturday, Sunday. If we had three months, we could do this."

CAG pasted on a benevolent smile. "While you, Lieutenant Zachery, are going to get it done in the aforementioned Friday, Saturday, and Sunday. Now, before you waste any more daylight, go figure out what you need."

Stretch shook his head. He was seeing the calendar. The next three days, he'd intended to spend some of each of those days with his family. Now, it was as if those days had been ripped off, crumpled, and tossed in the trash can.

"Stretch, goddammit," the skipper growled. "An aye, aye, sir would be in order long about right now."

"Sometimes, Skipper, *aye, aye, sirs* are one heck of a lot easier to say than to live up to."

"See, Little Lord," CAG said. "I told you he'd get it."

"Stretch," the skipper said, "you and Alice figure out what you need and let us know at 1300."

"I'll let you know right now, sir. We need"—he started tallying up numbers of pilots and what each would require—"twelve hours of China Lake range time. The whole range, not just the eastern slice. We need twenty-four hours of time in the TIAS simulator and the money to pay for it all. And I need to understand where the money is coming from and that everybody from you, Skipper, up to the chief of naval operations is okay with us spending it."

CAG started laughing. "See, Little Lord. I told you he'd get it."

Stretch got on the phone with CAG's ops O. CAG ops put him in touch with a commander on the staff of the three-star admiral responsible for naval aviation on the West Coast. The commander had had a heads-up the call was coming. Still, the commander was not pleased with the turd the Warhorse lieutenant had dropped in his lap.

The commander groused, "Fiscal year isn't half-over, and this is going to wipe out all the reserve budget we have, but I've been ordered to get you, Lieutenant Shithead, the funding. Never mind it means we'll run out of money before January is out. Getting us through the rest of the year will be a regular Mongolian goat rope. Here's a tip for you, Lieutenant. Shore duty sucks. Mainly because even a shithead lieutenant can make all kinds of grief for a guy two ranks above him."

"Sorry, sir. Thank you, sir."

"Lieutenant Shithead Zachery. I'm going to remember you."

The commander hung up.

Jon placed the phone back in the cradle as if it might blow up if he set it down too hard.

Mike looked up. He was working on a detailed flight plan for the next three days. "Made another lifelong friend, I see."

"Lieutenant Junior Grade Allison, you should concentrate on getting your menial task done, not ragging on your superior officer."

Mike laughed, clearly pleased he'd gotten his neighbor's goat.

Father God, who art in heaven, I've always got one foot in hot water and the other in a cold puddle of ridicule.

Jon stared at the black handset set on the black phone cradle. He expected to hear Teresa's *God won't give us anything we can't handle.*

Instead, he heard Pop: *Gittin' a worthwhile thing done, most times, ain't easy. Never expect it to be.*

Jon sucked in a lungful of air, held it for a moment, giving the fresh stuff time to change places with the stale and rancorous, and huffed out whatever the heck wanted to leave.

Jon's next two calls, to Ross Hill, the TIAS project engineer, and to Doyel Thomas, the man in charge of range time scheduling at China Lake, went considerably better. Both had been delighted to hear from Lieutenant Zachery again.

"No problem on the range," Mr. Thomas said. "Nothing scheduled for Friday, and, of course, nothing on the weekend. Here at the Lake, we're shore duty pukes, not See Dogs."

"Very funny, Mr. Thomas."

Doyel chuckled. "I told you to call me Doyel."

Jon replied, "I'll shoot you a message with the particulars as soon as I can get it typed up. Mr. Thomas."

Jon hung up.

"He called you Sea Dog?" Mike asked.

Jon glowered and held out his hand. "Schedule. Please."

Mike handed it over.

Mike had slated himself to fly with Blackey on Saturday.

"Blackey," Jon said.

"You still hate him for what he did at that party at the XO's house?"

Jon looked at Mike. "There's a heck of a lot more to it than that, and I don't hate him. What I really don't do is trust him. The rest of you in the squadron, if I get shot down over the North, I trust each one of you to do what you can to get me back. Shoot, I trust every pilot in the airwing to do that. Except him. If he and I were flying and I got bagged, he'd fly low over me while using his relief tube."

"Rooming with him on the ship must be interesting,"

Mike said. "I got him scheduled to fly with me. You want to fly with him?"

"Move him from the afternoon to the morning with you. I'll fly with him in the afternoon."

"Then we compare notes?"

"Affirm. Now let's go tell CAG and the skipper we got three months of work stuffed into as many days."

They gathered up the flight schedule.

"When're your folks going back?" Jon said.

"They're staying until Monday. They want to watch us fly off to the boat. What about Teresa's dad and the Conovers?"

"They're going Saturday."

They exited the office and started down the passageway.

Jon said, "I was going to take them out to the end of the runway tomorrow night so they could observe the bounces."

"One of the newbies could take them out to the airfield. Talk to Simp."

In the skipper's office, the same crew waited for them.

Jon and Mike tag-teamed describing the plan, the sims, the flights, and how many flights and sims each pilot would receive.

"The other thing, Skipper," Jon said. "When I talked to Range Control at China Lake, Mr. Thomas told me there was nothing scheduled for the range on Friday. Five months ago, when I got in a little trouble—"

"A little trouble!" CAG said. "You got into a gigantic shit pot of trouble."

"Aye, sir. When I got in big trouble, I learned some things about the range and billing for their services. They charge more when you reserve their entire airspace, even if there's no additional services or man-hours required. If you want the whole range, you pay extra to keep other projects from buying time too. But Mr. Thomas told me nobody else was scheduled on the range either tomorrow or the weekend. After I got out of big trouble, our Lemoore admiral called the China Lake admiral and worked out a deal to avoid the extra charges for using the whole range. If he can do it again, he'd save some of the West Coast naval aviation budget."

CAG said, "I'll call Lemoore admiral and ask him to call China Lake admiral. Then I'm going back to San Diego."

"CAG," Jon said, "when you're talking to the admiral, can you ask him if we can use the base passenger plane?"

"Little Lord," CAG said, "your Lieutenant Zachery, he's a regular piece of work, isn't he? Why don't you have Simp work that out?" To Jon, he said, "You have many talents, Lieutenant Zachery. One of them is blushing." CAG stood and walked out of the office.

"XO," the skipper said, "go. Worry him airborne. Stretch, we are convening a meeting in ten minutes for the guys going to China Lake. See if you can get your foot out of your mouth before then."

After Jon briefed the pilots on what to expect at China Lake, he went home for dinner. He was scheduled to fly on the first bounce hop of the evening. Simp assigned LTJG Walt Short to escort the Conovers and Velmers to the end of the runway so they could witness the carrier landing practice.

Tiny had been the grill maestro for the early part of the week. He'd returned to San Diego, so Mr. Velmer and Bart Conover took over the grill. Jon had invited Walt Short to eat with them.

The men were on the patio drinking beer and dodging meat smoke.

"Mr. Short, nice of you to be our escort tonight," Mr. Velmer said as he flipped the pork steaks and squinted.

"You're misters, Mr. V. and Mr. C. I'm Walt, or Stump. And I'm not doing this because I'm nice but because Stretch invited me to a home-cooked meal."

"The other reason is because the operations officer ordered him to," Stretch said.

"I was beginning to think I understood you guys and your call signs," Bart said. "Tiny is the biggest pilot in the airwing. Stretch is short. But now it seems this opposite thing applies to real names too. What are you, Walt, six three?"

"Only six two."

"Well, you're skinny as a rake handle," Mr. Velmer said. "I'm surprised you're not Fatso."

"Jed Newsome, Nose, assigns call signs in the Warhorses," Walt said. "God and Nose work in mysterious ways."

"Walt and Eli Banks are the two newest pilots in the squadron," Stretch said. "Nose said he didn't have time to think of proper call signs for Eli, so he is Bee."

"As in Newbie?" Bart said.

"As in one notch above nobody," Stretch said.

"He's just jealous he can't have a beer 'cause he's flying tonight. Meanwhile, I'll have another. Can I get you misters one?"

"I'd love to have another, Mr. Stump," Bart said.

"Me too," from Mr. Velmer.

And just like that, Jon felt like Mr. Nobody. *I can't have a beer.* A little thing, but that's all it took. In the squadron, show a hint of weakness or ooze a drop of blood in the water, and the piranhas swarm.

Jon sat on a folding lawn chair as the three men bonded. Together, they'd embark on an adventure in a couple of hours. Jon would be on a greater adventure, flying. He'd be a star of the show, but that didn't rank with the three of them doing a thing together. He also noted how Mr. Velmer and Bart Conover related.

The two of them had met in Kingsville, Texas, when the Zacherys lost Daniel. But the way they were getting along now, Mr. Velmer appeared to be talking to the ideal model for a son-in-law. Bart was a chemical engineer and a manager in a chemical plant in Kingsville. Mr. V. was a banker. Jon was just a junior navy officer and ex-enlisted man. Compared to the two of them in the business world, Jon was a nobody.

Walt returned with three beers and set two on the small table with the barbeque sauce. Mr. V. turned the meat. Mr. C. painted it.

"Looks good. Smells good." Walt, Stump, extended his bottle for a clink.

The meat turner and painter raised their bottles, clinked, chugged, and went back to work.

"Stump," Mr. Velmer said, "would you get the meat platter from the kitchen, please?"

Walt stepped inside and closed the sliding screen. Jon moped.

After his bounce hop, Jon drove home and entered their quarters at 9:15 p.m.

"Jon!" Carin Conover exulted. "It was such a special treat to see you, what do you call it? Bounce? I just had no idea of the precision required of carrier pilots. I mean, I saw almost fifty landings tonight, and I could not tell the difference between them. But Bee, he's an LSO under training, you know, he explained which landings earned four points, which three, and which ones only two."

"Yeah," Bart said. "Bee told us a trainee LSO observes hundreds of landings before he develops an eye that can tell the difference between a two, three, and four pass." His eyes sparkled. "The precision required all the way around. Man!"

To the baby in her arms, Carin Conover cooed, "Your daddy is a good stick. Yes he is."

Bart raised his beer bottle. "To the red-hot nasal radiator."

Carin hissed. "Jennifer and EJ are asleep."

Everybody was in the kitchen. One of those nights where the whole party wound up in the smallest—well, the bathroom was smallest—room in the house.

Mrs. Velmer pushed through the bodies to the fridge and took out a tray of ice cubes, then returned through the pack to the counter, where she poured Jon a drink.

Jon found Mr. Velmer's eyes on him. No expression on his in-law face. Sort of like Pop's, he mused. Except Pop's eyes were always downcast. At least Mr. Velmer looked at him. Through him might be more accurate.

Teresa stood next to her father. She looked tired, wiped out.

Jon said, "Teresa, are you all right?"

"I'm sorry," Teresa said, "but I have to lie down."

"Oh," Carin said, "Bart, we should go. I'm sure Jon is tired too."

"Carin," Mrs. Velmer said, "why don't you help me get Teresa and the baby settled. Teresa won't want you all to leave without saying goodbye. Once we have them in bed, the men can come in and whisper adieu."

The women exited into the hallway to the bedrooms.

Mr. V said, "Let's sit," picked up Jon's drink, and led the way through the dining room to the living room.

Bart sat on end of the sofa, Mr. V. the other.

"Mrs. Velmer told us you were burning your two-ended candle on three ends before the baptism," Bart said. "Looks like you're right back at it."

"Yeah. But that's what we have to do." Jon looked from one to the other. "It meant so much to us for you both to be here. It was asking a lot for you to come all this way for such a short time. Thank you."

"We were honored to be Ruthanne's godparents. And it was nice to see you in happy circumstances."

"As to burning the candle," Jon said. "Part of it is

due to how we are fighting this war. Very different from how Mr. Velmer fought World War II. He went into the Pacific theater and didn't come home until the war was won. Two and half years, you were gone, right?"

Mr. Velmer nodded.

"This is supposed to be a limited war. Maybe it's limited for Mr. McNamara, but it isn't for us."

A moribund silence filled the living room.

Until Mr. Velmer killed it. "Conversation killers. We used to call them discovering a floater in the punchbowl. Do you modern sailors still use that expression?"

Mrs. Velmer and Carin returned to the living room.

Mrs. V. stopped, put her hands on her hips, and turned to Carin. "Did you hear what my husband said? *Where two or three are gathered, there am I in the midst of you.* Which is true of course, except when the three are all men." She pointed to her husband. "You wash your mouth out with soap, then say goodbye to your daughter."

Carin handed Ruthanne over to Jon. "You too, Bart. We should go."

After the bedroom goodbyes and the departures, Jon put the baby in her cradle. Ruthanne emitted her tiny sigh, as if saying, "Finally I can stop being adorable and rest."

Jon knelt next to the bed.

"You're still coming home every night from China Lake?"

"Yes."

"Mmmm."

She did not like the expression *sleep tight*. Tight was tension. Good sleep was bereft of tension.

"Sleep loose, sweetheart," he said.

Her breathing said she already was.

Chapter 33

Monday. Back aboard the *Solomons* at 1550. Everyone was in their ready room chair for the 1600 APM (all pilots meeting). If the squadron ops O called the meeting, you arrived five minutes early. This APM CAG had called.

The skipper stood at the front of the room, eyes fixed on the back. He popped to attention and hollered, "Attention on deck."

"Seats!"

CAG's bellow caught most of the pilots at half rise. They flopped back onto their butts.

The skipper got out of his boss's way and sat in his chair.

CAG reached front and spun around. His eyes scanned the assembly.

Like a shark, Jon thought, who smelled blood and was looking for the bleeder. CAG's eyes swept the rows of seated pilots.

CAG was putting on an act, Stretch knew, but his Adam's apple bobbed when the eyes lit on him. Then, *Thank You, God,* the eyes moved on.

A TICKET TO HELL: ON OTHER MEN'S SINS

The airwing commander's glare stopped on LTJG Eli Banks. Call sign Bee. The junior pilot. Everybody turned around to look.

"Bee," CAG said.

Bee bolted to his feet.

"For Christ's sake. Sit. Down."

Bee sat.

CAG peered down at the deck, then raised up again with a pleasant smile on his face.

"Bee."

"Yes, sir?"

"Do you know what thumping is?"

"Uh—. Uh, I know who Thumper is."

Everybody laughed, even the skipper.

"Shut up!"

A clap of silence followed that CAG lightning bolt.

CAG's smile reappeared. "Blackey, explain thumping to Bee."

"Yes, sir. Thumping is when you sneak up behind another airplane and fly under him close and fast. Scare the shit out of the pilot."

CAG grimaced. "Troll, you explain it."

"Yes, sir. Say you see another plane orbiting over the ship. Maybe he's waiting for his turn to land. Maybe he's a tanker. Any rate, you, like Blackey said, sneak up behind him. Approach from his blind spot. You want to have at keleast a hundred knots of overtaking speed, and you pass real close underneath your thumpee. If you are close, the subsonic pressure wave around your airplane will jostle the thumpee and, in fact, make a sort of *thump* sound. The first time it happened to me, I peed my flight suit."

CAG nodded, smiled at Troll, and started walking down the aisle, as if that's what he'd come for: to make sure everybody knew the definition of thumping. At the last row of chairs, he stopped, and without turning around, he growled, "If any of you thumps another airplane inside or outside my airwing, you best hope you cut it too close and kill yourself, because if you survive, the hell I will visit on you will be worse than the one the devil would." Then CAG left.

"What the hell was that about?" Skippy said.

"Some of the Raider JOs are thumping each other," Nose said.

"Here's the rest of the story," the skipper said. "The Raider CO didn't know his guys were doing that stunt. But he arrived over the ship and saw one of his planes thump another. It wasn't hard to figure out the thumper and thumpee. Both those guys had thumped others before. Both are confined to quarters. They will go to mast this afternoon before CAG. Tomorrow they will appear before a *feenab* (Field Naval Aviator Advisory Board). If the feenab does not recommend pulling their wings, I will be the most surprised guy in this hemisphere. Not, you will sit on that board.

"One more thing. If CAG had witnessed the thumping, besides the feenab, the squadron CO, the XO, and the ops O would have been fired.

"Now, is there any question about CAG's message?"

The skipper's anger was every bit as ferocious as CAG's had been.

"Two more things. Nose, Bee needs a new call sign. Make it so."

"You mean Thumper, Skipper? Is that a good idea?"

"Make it so."

"Aye, sir."

"Stretch and Alice, come with me. CAG wants to talk Ironhand."

Teresa sat at the dining room table with Mother. Ruthanne sat on the table in her infant seat. Mother had her tea, Teresa orange juice. Ruthanne's eyes, as they did when she was clean, fed, and awake, roamed around, drinking in the wonders surrounding her.

Mother sipped her tea. "I woke this morning feeling like I suffered from postpartum depression. After all the festivities of last week, after finally getting to see Jon for a couple of hours yesterday afternoon, this morning was like falling off an emotional cliff. But it is impossible to feel depressed when I can look on our little angel. See, she's looking right at me."

Postpartum.

Teresa knew about postpartum, but she never appended depression onto it. Pregnancy was enduring and endeavoring to maintain a semblance of normalcy for her other children. She looked forward to it all being over with, only to discover her life had grown to depend on the particular kind of enduring and the particular kind of endeavoring as much as she depended on air to breathe. Through the nine months, there was the feeling,

If I can just get us through this ordeal, everything will be wonderful. The baby would have arrived. Things could go back to normal. But then you discovered pregnancy was the only normal you knew. And what you had was postpartum. And as long as she kept that D word out of it, there was hope that, *This too shall pass.*

EJ had his truck on the carpet in the living room. Mother spoke baby talk with little angel.

When Jon left, whether for seven months or seven days, *You take a part of me with you, It leaves a hole in me.* She'd written that to him.

He wrote back: *If I couldn't take some of you with me, I couldn't survive being apart from you.*

She judged it to be something he wrote late at night, when the rest of the ship slept, except for the ones on watch, of course, but not a thing he would write in the light of day. Although she'd considered writing about her judgment to him, she never did. It seemed somehow important that she not pop that bubble of Jon's belief. And, she admitted to herself, that bit of his faith was something she could cling to as well.

"You're our little angel, yes you are," Mother cooed.

Teresa smiled and shook her head. *Mother cooed! Will wonders never cease?*

"So, Zachery," CAG said, "summary of your weekend sojourn in beautiful downtown China Lake: every Warhorse pilot from Alice and senior to him is qualified and designated as an Ironhand lead."

CAG wanted him to say, "Except Blackey." But he didn't.

So, the airwing commander did, and, "Blackey is the best bomber, the best pilot on the ball, the best dog fighter, for an A-4 pilot that is. So why is it you don't think he can do this one job, Ironhand lead?"

"Oh, I think he can do it, sir. It's just that I don't trust him to do it."

"Alice. What do you think?"

"I agree with Stretch. Blackey can do the job."

"What about trusting him?"

"Well, sir, Stretch doesn't trust him. I'm wary."

"Wary!" CAG said. "Jesus Christ, Little Lord. You have a wary lieutenant junior grade in the Warhorses. If Alice wasn't Deputy Dog Ironhander, I'd move him to the Raiders where that bunch of shithead thumpers could use an infusion of wariness."

CAG swiveled his head toward Stretch. It reminded him of 1966 and the after gun mount on his destroyer training on a target.

"What if I ordered you to designate him an Ironhand lead?"

"I'd say, 'Aye, sir,' and do it."

"But?"

"But I would ask the skipper to put me on as Blackey's wingman every time he flew lead."

"Because you're wary and the rest of your babe-in-woods-JO squadron mates aren't?"

Stretch's anger ratcheted up a notch. CAG sounded like Blackey, before his wife divorced him. What the heck was the guy doing?

"Well?" CAG said.

Stretch looked him in the eye. "Because I'm wary, sir, and the rest of my babe-in-the-woods-JO squadron mates aren't."

For a moment, Stretch thought CAG was going to stare him down, like Commander Fuller did—what was it? A hundred years ago? In the next moment, he wondered if a kiloton explosion might be brewing up behind the man's dark eyes.

CAG sat back and glanced at the skipper. Stretch felt as if a giant hand had locked him on his chair, but now it let go, and the skipper's boss had become full of benevolence instead of filled with senior-officer menace.

"One more thing, Stretch. Tell me about this defensive ACM—air combat maneuvering—program of yours."

"It's Troll's idea, sir."

"Okay, it's Troll's. Spill it."

"I think every ACM hop I've flown has been offensive. Try to get into position to shoot the other guy down. That's what I was trying to do at Fallon. I talked to Troll, and he says the likelihood of an A-4 shooting down a MiG over North Vietnam is zip. If we get jumped by MiGs, he says we should maneuver to keep the enemy in sight. We should keep our speed up by trading altitude for knots. The way an encounter is likely to go is a gaggle of MiGs will hit an Alpha strike going fast and shooting as they blow through us. They might get clear, turn around, and hit us with a second slashing attack. According to Troll, Tiny, and air force intel wienies, the bad guys will head for the barn after a second attack. If the bad guys do stick around, our fighters ought to be in position to pull our bacon out of the fire."

"You do know an A-4 guy from Lemoore did get a MiG with a Zuni rocket."

"Yes, sir. Troll says if you put enough blind sows in a forest of oak trees, one of them is bound to find an acorn."

The skipper piped in. "All my pilots have gotten a defensive ACM brief from Troll, and we've all flown against him. What we are going to do during this time at sea is all of us will fly against Blackey. If we can hold him off through two attacks, we figure we can handle a MiG."

"Has anyone tried this defensive ACM program against him yet?" CAG said.

"Yes, sir." The skipper again. "Half of us. And half of the half held Blackey off, kept him from getting into position to shoot the target guy down with guns. The problem with Blackey though is, after the hop, he can't explain to the target guy what he did wrong. Now we have Troll fly above the fray, keep score, and give the target guy pointers after the hop."

"This working for you?"

"It's working, CAG."

CAG nodded. "Okay, Little Lord. This is what I want you to do. On our second practice Alpha strike, let Blackey fly Ironhand one. Stretch, Ironhand two. After the hop, the four of us will get together again right here and see what we've learned. And two days from now, I'd like to fly one of your planes against Blackey."

The dishes were done, the children were blessedly asleep, and Teresa intended to get out her stationery and write to Jon.

Mother, however, had another idea, and a martini in one hand and a glass of Pepsi in the other.

Annoying. She never sat with Jon the times he wanted a drink at home. Drinking was his business that he knew she disapproved of. Mother had to know that. Of course, other people's disapproval, at times, seemed to be her motivation to annoy them more, to make them disapprove more.

But, Teresa Velmer Zachery, you owe your mother a few minutes of your precious time since she's given you months of hers.

She reminded herself not to sigh and accepted the Pepsi.

Mother sat and sipped. "You want to write to him. I can feel your stationery box pull at you. Thank you for sitting with me. Jon and I sat like this a couple of times. I found out I never really knew the man you married. I disapproved of him from when you two started dating, and I never ... moved past that. Talking with Sarah Fant, I knew his CO thinks highly of Jon. When I sat and talked to him, with my mind open to see beyond my prejudice, I found not only a good man but an exceptional one."

Another sip. "And it made me think about how well I know you. How I don't know you well at all. It took me seeing you in the Warhorse wives and each of them an exceptional woman, and that you fit right in with them. This really is an exceptional group. Amy Allison told me about your wives' group from last cruise and that there wasn't this element of closeness. Sarah Fant told me that once before, they were in a squadron that was like this. The wives were close. The men were close. She said it was a blessing to have experienced it once, but to find it again, this magical assembly of personalities that blend so perfectly, well, she said, 'I don't know what comes beyond a blessing, but there must be something between a blessing on earth and heaven. And that's what the Warhorses are.' And what I know, Teresa, is that you contribute every bit as much to making your group so exceptional as does Sarah and each of the others. It was a blessing for me to be able to come out here and spend enough time to see that."

Teresa didn't cry, but her eyes watered. She started pushing herself to her feet, grimaced, and eased the rest of the way to her feet. *So easy to forget the incision.*

"Don't make me spill my martini," Mother said.

Teresa restrained her hug.

"You're going to put our conversation in your letter, aren't you?"

Mother did her best to look annoyed when her daughter confessed, "Yes."

After a second martini, Mother went to bed in the children's room. Teresa started her letter.

She was tired, but the talk with her Mother had been so … unexpected, so … longed for. It had been unobtainable right up until it fell in her lap.

Jon had told Mother about Teresa's mysteries of the

rosary. She wanted to hear about the Thank You, God mysteries. Teresa described them as Mother drank her second martini.

A martini with an olive and a rosary! Who but Mother?

She closed her letter with a prayer for Jon and all his shipmates on the *Solomons: Father God, who art in heaven, please don't let anybody die tonight.*

Chapter 34

Stretch grabbed chow with Alice in the dirty shirt dining room and then headed to his room.

Blackey lay atop his bunk in his flight suit, reading a paperback. He didn't look up or say anything.

Stretch stepped to his desk, pulled the lid down, and sat. He intended to start his letter to Teresa. That night, he had to log two landings to requalify for night carrier ops, but that was hours away.

For a moment, he listened to the noises coming from the hangar deck above. Sailors dropped equipment and threw gear, and it banged and clattered. At times, it seemed sailors engaged in a daylong contest to see who could make the most noise, as if, at the end of the day, there'd be an announcement, "The winner of today's noise contest is Seaman Smedlap from the Hangar Deck Division." Seaman Smedlap and his buddies were good kids. They worked sixteen- to eighteen-hour days at a sweaty, greasy job. Good kids. They made the carrier go every bit as much as the propellers did.

Stretch pushed them gently aside, reached for quiet, and pulled some to the center of his mind. Teresa came to him. He watched her take care of their children, interact with her mother, pray her rosary with her Thank You, God mysteries as she looked down on Ruthanne in her cradle. He watched her be herself in her body, her mind, and her soul. The need to love all three of her in ink rose in him.

He wrote the salutation. And stopped. He felt a chill on his back. The ventilation vent wasn't blowing on him. Blackey. He lay on his bottom bunk as before, reading, acting as if he was unaware of his roommate. The chill came from him, as if he was a cold soaked stone wall.

Stretch turned back to his letter, and the chill on his back intensified. He looked at his hand holding his pen.

Decision time. Stay and don't let Blackey win. Or forfeit and get back to loving Teresa. Loving her won.

In the ready room, he got a mug of coffee and wrote himself to some place less warm and happy than before but enough so that Teresa was there.

Stretch was on the second go. He would hot-seat into Blackey's plane, 514.

The flight deck at night was a high pucker factor place, and especially during night carrier landing qualifications. The eerie red light, a dozen planes parked with their engines screaming, and overall, a palpable sense of urgency spurred a pilot to find his plane, switch with the guy inside, and launch into the pattern to log his landings.

Stretch gave a thumbs-up to Blackey in the cockpit.

He climbed out. At the foot of the ladder, he didn't say anything about the plane. Hot-seating, a departing pilot always said something. "Good bird." "Radio's scratchy, but it works." Something. But not Blackey. Not Blackey to Stretch.

Stretch climbed up the ladder and into the cockpit. The first thing he did was to check the radar scope. Brackets secured it into the instrument panel. Bolts tight. He always checked the scope, but tonight, with Blackey behaving as he did, he paid extra attention.

Takeoff checklist. Thumbs-up to his plane captain. The flight deck taxi director signaled, "Hold the brakes."

That meant all the tie-down chains were being removed and the chocks would be pulled. Only the brakes fixed LT Jon Zachery to his spot. His brain tried to tell him his plane was sitting on a sheet of ice and that, any minute, the wind would start sliding him toward the edge of the deck and over into the deep, dark water.

A fleeting thing, an ephemeron. But a healthy dose of pure fear he had to deal with every time he manned up at night after a layoff of a week or two. Five in this case.

On a flight deck, though, once they decided your plane needed to move to the cat, you moved without any diddle-dorking around.

In the universe of darkness, there was only the taxi director's wands. The wands signaled, come forward, come forward, turn right, come forward, come forward, come forward. Easy now, *eeeeasy* forward. Stop.

He stopped and smiled at himself inside his oxygen mask. *You're such a sissy, Zachery.*

Clunks from outside seemed to make 514 shudder, as

if she dreaded the cat shot as much as he did. A spurt of self-preservation supplanted his moment of self-ridicule. He anticipated the feeling of utter helplessness, utter dependence on the catapult to hurl him into the ink and 514 to sustain the force of it and not fail in any of her vital parts, and during that second of travel down the cat track, to know if something did fail, if something did go wrong, there was not one bloody thing he could do about it—until the force of the cat shot let go of him. Then he would have one second to react, to save his life.

New wand signals. *Feet off the brakes. Power full.*

Inside the cockpit, check the gauges. Good. Cycle the control stick. Free of binding. Turn the lights on.

Wait! Did I check the radar scope?

The catapult didn't hear and didn't care. Five fourteen squatted momentarily, and then a giant boot kicked the plane down the track and pressed Stretch back against his seat. And this was the worst kind of cat shot. His plane was light loaded, no ordnance, and not much gas. That meant the shot would feel like it wasn't powerful enough. Ninety-nine percent of the time, a pilot got a heavy load shot. The light ones were just scary.

Then the cat forces let go, and there was another moment of question. *Are we flying?* But no time to wait for an answer. Set the nose up attitude. Airspeed: good. Climbing, not descending. Also good. And an answer to, Are we flying? His left hand came off the cat grip and reached for the landing gear handle. *No.* The gear had to stay extended. He and 514 had to turn directly into the landing pattern.

He keyed the mike. "Five fourteen airborne."

"Five fourteen, level at twelve hundred feet and turn downwind."

"Five fourteen."

The next thing was to level off when his body and soul voted to climb, climb, climb. Get farther away from the water and death.

But he got the plane leveled at fourteen hundred feet and descended back to twelve in the turn. That's when he wondered if he had screamed during the cat shot. All he ever knew for sure was he hadn't keyed the mike and screamed over the radio. He wondered what new call sign that would earn him. But then, such thoughts no longer had a place inside the box of rocks that passed for his brain.

He completed his two landings and turned 514 over to Thumper.

In the ready room, he returned to his letter as he waited for the LSOs to give him his threes for grades, which is what they gave him, and then debriefed the other Warhorse pilots.

Blackey sat across the aisle from Stretch, working on the flight schedule for tomorrow.

Stretch took his letter back to the stateroom and sat at his desk with the overhead light off. He took himself back to where he'd been before Blackey frosted him, and he found Teresa in body, mind, and soul, and he loved the three of her in ink. As he wrote, carrier sounds came to him from far, far away.

Teresa X-ed off the days on her calendar. She took care of her children and, with a lot of help from Mother, took care of the house. She wrote to Jon.

On Friday, Teresa and Amy shopped together at the commissary while Mrs. Velmer babysat. Sunday, Ruthanne attended Mass with the family and behaved like a Christian—thank You, Father God, in heaven—not a howling heathen. Jon's first letter came on Monday. It had taken a week to arrive.

"It takes a week for a letter to get here from the Tonkin Gulf," Teresa told her mother. "Jon's off the coast of California. Why does it take so long?"

"The US Navy and US Post Office are run by men."

Teresa gazed at the letter in her hands.

"Aren't you going to open it?"

Teresa shook her head. "Tonight, when I can write him back."

Mother rolled her eyes. "When your father was in the Pacific and I got a letter, I tore it open at the mailbox and read it a half dozen more times before I went to bed."

"I'll read it tonight."

Another eye roll, and Mother invited EJ to join her on the sofa. They read a book, then they acted out the story. Sometimes she made up bits of costume to wear.

When Teresa gave Ruthanne her bath in the sink, with EJ on his chair watching, the umbilical scab came off. EJ was so excited to discover his sister had a belly button, as if she hadn't been a regular kid without one.

That afternoon, Teresa started her letter to Jon. The belly button had to go in, but she wanted to deal with it before she read and responded to his.

A TICKET TO HELL: ON OTHER MEN'S SINS

Stretch and Alice sat on folding chairs in the skipper's stateroom.

"Blackey led the Ironhand on the Alpha strike today. How'd he do?" the skipper said.

"I couldn't fault him for anything he did or didn't do on the mission. He played the game."

"But you still don't trust him?"

"No, sir. I do not. On these practice Alphas, there's always CAG or one of you squadron COs flying above and observing."

"What about you, Alice?"

"I'm a JG, Skipper. I think it twists his skivvies in a knot to fly as my wingman. But, on the practice alphas, like Stretch says, he plays the game."

"What you're worried about is what'll he'll do in combat, right?"

"There's a lot of Blackey he doesn't let anyone see," Stretch said. "The rest of the guys, you talk to them, you fly with them, you get a sense of what you can count on from them. With him though, I'm left with what's in that hidden part? What will he do over the North in a tense situation? Will he decide he knows better than us dumb butts you mistakenly put in charge of him, God's gift to naval aviation?"

The skipper switched his gaze to Alice.

"I worry about him too, Skipper. Probably not as much as Stretch, but then I don't have to live with him. Every other pilot in the squadron, I trust them with no ifs, ands, or buts. Blackey, in the matter of trusting him, I got ifs, ands, and buts all working."

"Okay, so no Ironhand lead," the skipper said. "What about Wild Card? Seems like that might be an ideal fit."

Stretch looked at Alice. He answered, "On the Wild Card, a key thing is the TIAS scope. You have to scan outside the cockpit, of course. But you have to make yourself check the TIAS scope too. Frequently. That's the whole purpose of Wild Card. To be off to the side with SAM sites focused on the Alpha strike. The only way you'll see an indication of a SAM site is on the TIAS scope. If the site isn't looking at your aircraft, none of your other threat warning systems will alert. And he's badmouthed loud and long the idea of looking inside the cockpit when we're over the North. I don't trust him at all to do the Wild Card mission."

"My department heads?" the skipper said.

"Not is good for Ironhand lead and Wild Card. After Alice, he and Troll are solid. Simp and EC, Ironhand lead but not Wild Card. And I'm ready to designate Nose for Wild Card."

"Okay," the skipper said. "The defensive ACM training we've done has paid off too. Haven't lost anyone to the bad guys. So far."

The last training alpha was tomorrow.

Stretch rapped his head, as did Alice.

The skipper grinned, but then he rapped on his head also.

A cat fired, and the force of it juddered through the ship. The other cat fired.

The ship and airwing were in hour twenty of their forty-eight-hour final graduation exercise to prove they were ready to deploy. Noon tomorrow, they had to hang on till then.

During one of these battle drills, sleep was precious as life itself. But it was hard to come by. Aircraft had to be maintained in alert condition to respond to pop-up threats dreamed up by the devil referees.

"Grab some shut-eye if you can," the skipper said.

The navy meant the final battle problem prior to deployment to be a challenge. Could a ship and wing hack it when the going got tough—really tough? That was the question wanting an answer. A second question was, "Can you do it without killing anyone? Replacements are expensive, you know." The battle problem was an endurance contest as well as a combat capability problem for the ship and airwing. The problem was structured to require the carrier to employ every offensive and defensive capability it had to survive and win.

Stretch was on an alert fifteen watch from midnight to 0400. The good thing was pilots stood alert fifteen—meaning his flight had to launch within fifteen minutes after receiving notice of a threat—in the ready room, in their chairs, and asleep in their flight gear. Alert five required the pilot to be in the cockpit and strapped in.

Stretch was lead of a four-plane antisurface ship mission: prevent an enemy surface combatant from getting within seventy-five miles—surface-to-surface missile launch range—by flying over the enemy vessel at low altitude and high speed and simulating dropping retarded bombs on it and blowing it to smithereens. His wingman was Thumper. The mission was going to be as tough a challenge as any he'd faced, and Stretch worried about subjecting the newest newbie to the demands. Night launch, rain showers and squalls forecast, radio silence, low-altitude, high-speed attack over water. At 2330, when he briefed the flight, Stretch asked Not, who was to man a spare aircraft, and Troll, assigned to fly number three, what they thought about Thumper going on the mission.

Not said, "Sooner or later, newbie's gotta learn to hack it."

"Yeah," Troll said, "but is tonight too soon with too much?"

Thumper piped in, "Hey! You assholes are talking about me like I ain't here."

They continued to talk about him like he wasn't there. Then, the three of them, with Thumper left behind, ascended to the flight deck to check on the conditions.

"Black-assed son-of-bitch," from Not.

Wind whistled around the island. Metal parts of the hoists and pulleys that ran signal flags up the mast rattled and clanged. The deck looked wet.

"Launch the alert five. Launch the alert five," blared from the flight deck announcing system.

Engine start cart turbines whined to life. The ship started a turn to bring the wind down the flight deck and heeled significantly. The engines of the two F-8s positioned on cats cranked up.

"I've seen enough," Stretch said.

Back in the ready room, Stretch said, "Okay, Thumper. You're on the schedule. If we're called away, you go with us to the flight deck, but if conditions are still like we just saw—"

"Or worse," Not cut in.

"Or worse, I'll tell you outside the island that you're scrubbed. You will not argue with me. You will turn around and come back down here. Understand?"

"Shit," Thumper said.

Not grinned. "He *un der stands* and said so with a word that has only one syllable."

At 0330, as the next five Warhorse pilots showed up to assume the alert fifteen, "Launch the alert fifteen was called away." The new crew hadn't briefed yet, so Stretch and his flight scrambled to the flight deck, where the abysmal weather conditions sent disappointed Thumper back down to the ready room as the others flew the mission.

At 0515, Stretch and his flight landed, debriefed, and returned to the ready room.

A sullen Thumper sat in the last row of chairs.

Not told him, "If I was you, which I ain't, I'd kiss Stretch's ass. He saved your life."

"Shit," Thumper said.

Chapter 35

Teresa and her mother stood in front of the calendar.

"Jon comes home today. I would have circled the date," Mother said.

"Some people are superstitious. We've had so many people bring food or been here to help in other ways. I didn't want to offend any of them."

"You mean if you circled the date, it would jinx it? Like the squadron wouldn't be able to come home?"

"Some pilots and some of the wives believe voicing happiness over an anticipated favorable outcome is a sure way to dash that happy feeling. That's how Jon puts it."

"You and Jon aren't superstitious."

Mother said things with assurance. Most people would turn that sentence into an interrogative.

"Haven't you seen Jon rap on his head when someone says something … jinx worthy?"

"No."

"He does, though, and sometimes he'll say, 'Best wood around.'"

"Of course, *he'd* be superstitious. Most country bumpkins are."

Mother left the kitchen, entered the bathroom, and closed the door. She'd gotten in the last word, as usual. What she'd said about Jon. A joke? Surely. Jon had written about the conversations he'd had with her. They were getting along well. Still, there was a snippy tone to her words. *Jon's a country bumpkin. I can forgive him for primitive behavior. You, Teresa, are my daughter. I expect better from you.*

Father God in heaven. Will I ever grow out of my mommy issues?

Ruthanne announced, from the bedroom, she had issues of her own that a mommy needed to deal with.

The alarm went off at 0730. Stretch smashed it off and draped an arm over his eyes.

For him, glorious sleep tasted better than glorious food. Teresa loved the music from *Oliver*.

Good morning, sweetheart.

Deep breath in. Huff it out. Throw the blanket back. Swing his feet over the end of his top bunk and climb down. Blackey was not in the bottom bunk. Stretch smiled.

Don shower shoes. Flip-flop to the head. Shave and shower. The flight suit would do for another day. Breakfast in the dirty shirt.

The ready room looked like a morgue body dump following a major catastrophe. Comatose forms slouched in the seats, eyes closed. Some had mouths gaped open. Thumper's mouth and eyes were open. In the junior officer bunkroom, Nose discovered him sleeping with his eyes open and claimed zombies slept that way, and he changed Thumper's call sign. Nose also confiscated a corner of the whiteboard in front of the ready room and labeled it: "LTJG Bank's weekly call sign. DO NOT ERASE." So, Zombie he was. For another day or so.

LT Howie Wisdom, assistant maintenance officer, Wiz, sat behind the duty desk, awake but with raccoon eyes and fatigue in the process of melting his face off. Every pilot had been needed to fly to meet the requirements of the battle problem. Every pilot had flown once during the night. The skipper and Blackey had both logged two night hops.

The clock by the SDO desk read 0815. The battle problem was scheduled to end at noon.

The force of a cat shot juddered through the ship. On the closed-circuit TV, an F-8 climbed away and disappeared into the goop. Stretch counted: *One potato, two potato*. At fifteen potato, the other cat fired. In bad weather, the ship had to allow an interval between launching planes.

The battle problem would most likely require the launching of alert birds right up to the crack of noon. One other major test was a virtual certainty, and that was the launch of an Alpha strike.

On the whiteboard, Wiz had grease-penciled ALPHA STRIKE. Beneath the title, he'd listed four pilots to fly as bombers and six as Ironhand, two of which were Wild Cards.

"Launch the alert fifteen. Launch the alert fifteen," blared over the ship's announcing system.

Not, Nose, and Nooner woke with a spasm of flailing arms. Not jumped up, found Nooner, grabbed his flight suit, pulled him to his feet, got in his face, and snarled, "Can you handle this?"

"I can handle it. I can handle it," Nooner said.

"Touch me, and I'll knock you on your ass," Nose said.

Beefy Not laughed at skinny Nose. "Grab your gear. Let's go."

The three of them snatched their helmet bags from hooks in the overhead and headed for the rear of the ready room.

Stretch considered Nose a solid airplane driver. Nooner was maturing, almost solid. Even the newest guys, Mudder and Thump-Zombie had developed considerably during the at-sea period. The weather last night and today added value to the training, as long as they lived through it. Stretch rapped his head.

"Stretch," Wiz said, "Not was flying Ironhand on the Alpha strike. I'm plugging you into his slot."

Wiz was doing a great job managing a limited number of pilots for a fluid and demanding set of requirements. For a ground pounder anyway.

"Here's the briefing cards for the Alpha strike," Wiz said. "Look them over but then put them right back on the edge of my desk."

Stretch nodded. The way the battle problem was going, he might get plugged into another alert and launch on that, and Wiz would have to find someone else to fill the Ironhand slot.

The battle problem levied an insatiable appetite for

up airplanes on the squadron maintenance department. If a plane returned from a mission in down status, it had to be returned to up status PDQ.

The good side of the ledger, and there was one, was that the end of the dadburned battle problem was in sight. The sun was up. The ship had moved closer to the coast of San Diego, where the weather was better. Now, a ceiling of a thousand feet hung over the carrier versus the three hundred footer Stretch had landed through last night.

Last night. Take off directly into the goop. Climb. Check in with the E-1 controller. Follow his vectors. Descend to low level. Accelerate to max speed. Attack the target and come back to the ship. Descend under radar control behind the ship. The attack on the simulated bad guy ships, right at the edge of what he could handle. *Finally,* he'd thought, *something I can handle easily. Night carrier landing.*

The controller said, "On glide slope. On glide slope. Onnnnn glide slope. Right one." Meaning turn right one precise degree. Just one. Then, "On glide slope. Right another one. Onnnn glide slope."

He was flying a perfect approach. *Perfect!*

Then he broke out of the goop, and there was the carrier, and it was *so* close. It almost didn't register that he was set up perfectly. The ball in the center, right on the landing area center line. A voice in his head screamed, *Do something! We're going to die!*

But the other voice shouted, "Ball, ball, ball." He was so close to landing the only thing that mattered was keeping the ball dead centered. Either his hands listened to the second voice or God landed the airplane. After he was parked and tied down, he had to sit a moment and gather the strength to climb out of the cockpit.

He'd handled it. But so had the new guys. *Thank You, Father God in heaven.*

Sitting in the ready room now, and thinking about it, the squadron new guys had handled some tough challenges through the night. He remembered RT telling him, "War is hell. For newbies, war is the ninth level of hell." Stretch had a corollary for RT's thesis. Training for war is hell too.

At 1000, the skipper came into the ready room and spoke to Wiz.

The ship had to deal with a pop-up group of four enemy surface ships steaming in formation and getting close to missile launch range against *Solomons*. Each A-4 squadron had to launch four planes to counter the threat.

The skipper picked EC, Stretch, and Skippy for the antisurface ship mission.

After briefing the mission but before leaving for the flight deck, the skipper told Wiz that he was sure the Alpha strike would still launch, that when it did, the Warhorses would be short of airplanes but to make sure that the Ironhand mission plane requirements were met before the bomber plane requirements.

Urinate, hustle to the flight deck, quick preflight, and climb in. Start engine. Cat shot. Rendezvous on EC. Attack enemy boats.

Orbit overhead while *Solomons* launched the Alpha strike. The communist battle problem refs didn't call for the strike until the very last minute, which would draw out the exercise two more hours. Stretch thought the squadron had had a good chance to be home by 1600. Now they'd be lucky to get there in time for dinner.

Then the skipper led his flight down to land.

Teresa was on the sofa nursing Ruthanne. The phone rang. Hopefully it was someone with news of when the guys would fly in. She heard Mother answer. Then say, "Oh, Maryann, any news?"

Teresa arranged the spit-up towel on her shoulder and the baby on it and patted her back. Ruthanne burped. "Good girl."

Mother stood in the doorway from the kitchen. The look on her face inspired a hand with fingers of ice to grip Teresa's heart.

"Jon?"

Consternation flitted over Mother's face for a moment. Then she shook her head. "No. No, Jon's okay. But there was a collision. Blackey and Troll. They rescued Blackey, but they haven't found Troll."

Profound relief flooded through Teresa's veins and bones.

Thank you, Father God, who art in heaven. It wasn't Jon or Mike. Or a newbie.

Then guilt and grief blew up inside Teresa with a silent *boom*, which emptied her of relief.

Troll. Homely, self-effacing, good-hearted. She wondered if he had family. She wondered why she didn't know, why she didn't know him any better.

A TICKET TO HELL: ON OTHER MEN'S SINS

Why did this happen? I didn't circle the date.

At 1325, Stretch, along with the other Warhorse pilots not on the Alpha strike, sat in the ready room and watched the flight deck TV as the planes landed. Once the strike was all back aboard, the planes would be fueled and prepped for the flyoff to Lemoore. Home in time for dinner. No sweat.

The phone rang on the SDO desk. Wiz answered, "Ready five, sir." He listened a moment, then said, "Shit," and hung up.

The skipper was in his front-row seat.

"A midair," Wiz said. "Blackey and Troll. A plane from the Raiders saw the planes go down. Only saw one chute. He's over the crash site and talking to Blackey on his survival radio. Hasn't heard from Troll. Blackey said Troll thumped him."

Nose sat forward in his chair. "Troll thumped Blackey? Bullshit. If there was any thumping done, Blackey did the thumping."

"Nose," the skipper snapped. "Shut up. I don't want rumors to start flying. We need facts." The skipper turned. "EC, you're lead accident investigator. Collect statements from everyone on the Alpha strike. Did anyone see the mishap? Did anyone see how Blackey and Troll flew during the mission? We need to move on this. A lot of the ready rooms probably did the flight sched like Wiz did, on the grease board. So grab the guys who are here and start sending them around to ready rooms and collecting statements. Stretch, you're EC's deputy. Not, get with the ship's ops O. Ask him to see if any of the radar controllers heard anything or saw anything out of the ordinary on radar. You know. Like one airplane overtaking another at high speed. Like would happen in a thumping. EC, have your flight deck CPO collect statements from the plane captains and maintenance troops on the flight deck when the Alpha launched."

The skipper stood statue still.

Stretch glanced around. All the pilots sat frozen to their chairs. He thought they were all like him, a spring wound tight. The skipper glared at EC.

EC jumped to his feet. "Stretch, assign guys to visit the other ready rooms. Give them some suggested questions to ask. Skippy, you gather statements from our pilots as soon as they get back. Hop to it."

The squadron had nine up airplanes left and thirteen pilots. At 1800, the ship started launching aircraft to fly to home bases. Blackey was still in sick bay. Stretch and Skippy were occupied with sorting and organizing all the statements that had been collected. They did not fly off. Neither did the junior pilot, Zombie.

The ship was home-ported at Alameda Naval Air Station on San Francisco Bay. It wouldn't arrive until tomorrow, Saturday morning.

Sorry, Teresa. I won't make it home tonight.

Stretch and Skippy sat side by side in two chairs in

the front row turned to face aft. Each of the four chairs they faced in the second row held a stack of paper on the seats.

Stretch held up a sheet of paper. "Skippy, this is good."

Skippy, black crew cut, thin, pale, serious face. Deep, dark eyes that shallowed and softened. His face crinkled into a smile, pleased at the praise.

He'd listed:

Possible accident causes:

1. Material failure in 514, Stoll's airplane, or 511, Blackey's.

    A review of the maintenance records of both aircraft showed only routine failures and no pattern of repeated failures in a critical system. Blackey said he'd had no system failures.

    Assessment: material failure possible but highly unlikely.

2. Bomb damage to 511 or 514. For the mission, both aircraft carried live five-hundred-pound bombs.

    Neither CAG, who flew as an observer, nor any other Alpha strike pilot observed anyone pull out low. Blackey, Ironhand two, reported his lead, Troll, had dropped his bombs at five thousand feet and bottomed out of his pull up at thirty-five hundred. Above the fragmentation pattern of the

> bombs. Assessment: damage from live bomb frags possible but highly unlikely.
3. Pilot error.

"You got some ideas about this last category?" Stretch said.

"Some."

"Well, keep going on it. I'm going to box up Stoll's stuff from his room. Otherwise someone will have to come back up to the ship during the Christmas leave period to do it."

In Stoll's safe, Stretch found a letter, from a Carolyn with a last name he didn't recognize. In the letter, she said she'd arranged her schedule during the heavy Christmas flying period so she could spend some time "with you, Robert."

Stretch sat back. Could Carolyn be Carolyn White, Blackey's ex? Had she taken her maiden name back after the divorce? This Carolyn and Troll, Blackey's former best bud, were going to spend time together. Did Blackey know?

Another thought. Blackey had been the ACM rooster atop the dung heap. He could shoot anybody down. Then Troll taught the Warhorse pilots his defensive ACM tactic, and most of them held Blackey off. He could no longer shoot them down.

Blackey knew about Troll and Carolyn. The evidence wasn't circumstantial. It was conjectural. Still, Stretch was sure. And Blackey was a hater. Nothing conjectural about that.

Chapter 36

Stretch sat on a back-row chair in the ship's chapel. In one way, it seemed like he'd been there yesterday trying to get the murder of Amos Kane into some form he could live with. In another way, Amos had been ten years ago.

Blackey had killed Troll. He knew it for truth.

Troll had been flight lead; Blackey his wingman. Blackey said after they bombed the target, which he hit and Troll missed, Troll had passed the lead to him. Then thirty-five miles from the ship, Troll had run into him. "Obviously trying to thump me, only he got a little close."

A little close.

No one else saw or heard anything to contradict Blackey's version. There was no evidence, no proof that Blackey's story was a lie. Except Stretch knew it to be one.

Blackey had drawn Troll to him, and he had captured Carolyn's heart, but Stretch had seen only repellent, cold, dark evil in the guy.

In the Warhorses, Stretch was Mr. Ironhand.

Blackey was used to being considered the best stick. And Ironhand was the most important mission for the squadron. The others listened to LT Zachery, not LT White. And Blackey didn't dislike Stretch for his role; *he hated me.*

Stretch could see what the skipper would say, though. "Look, Stretch, I know Blackey is an asshole, but what you're saying, there has to be more than your gut to make him a murderer. Everybody knows you two don't like each other. Are you thinking clearly about this?"

The most important thing was, Who would Blackey hate next?

Stretch closed his eyes. He was tired, drained, exhausted, but there was no sleep in his weariness. He kept his eyes closed, and stillness began to envelop him.

After a time, the thought came to him: *The only one Blackey hates is me.*

Stretch opened his eyes and sat up straight. Blackey wasn't a danger to the others.

That made all the difference in the world.

Thank You, Father God in heaven.

Stretch returned to the ready room at 0100. Skippy was still there, scribbling away on a yellow pad of paper. He stabbed a period on the page and said, "Done, by God. I'm going to bed. I may still be asleep in the JOB in January when we pull out for the Tonkin Gulf."

Zombie slept leaning forward on the SDO desk. A puddle of drool stained the green desk blotter pad under his face. Stretch woke him and sent him to bed. Then pushing the blotter aside, he started editing Skippy's accident report. The kid had done well. Stretch finished five pages and called the duty yeoman to pick up the pages he had finished and to type them up into accident report format. Then he continued revising the pile of paper Skippy had created.

At 0400, Stretch turned the final pages over to the yeoman.

It was so like the end of last deployment. Stretch had worked late on the report of what had happened to AB and Skunk. Now he had a report dealing with Blackey and Troll.

The YN came and left with the last pages. Stretch took a plastic trash bag, put Carolyn's letter in it along with a piece of scrap metal, and tossed the bag over the fantail.

Troll, I don't know if I'm doing the right thing. Maybe not. It just seems to me the squadron doesn't need this right now. I'll tell you, though, I won't forget you.

The lights of the California coast shone off the starboard side. Waves splashed and whooshed against the side as the big ship plowed through the darkness toward the Golden Gate Bridge.

The ship was almost home. He'd be back in Lemoore via navy transport aircraft by noon, deliver the accident report to EC, and be home by 1300.

The accident report might eat up some of Sunday and Monday. A memorial service for Troll was scheduled for Tuesday.

But then he had two weeks of leave. He would take peace on earth from his family, and thought, *Please, God, help me give them something too. So I'm not just a taker.*

As Stretch stood on the fantail in the dark, he thought, *War is hell, and so is Blackey.*

But that was for 1972. He shoved Blackey out of his head.

I'll be home for Christmas. For real.
Please, Father God in heaven?
Knock on wood.

VA-92 Warhorse Officers 1970-71 Deployment

Operations Officer	LCDR Dave Clark	AB
	Spouse, Sybil	
Maintenance Officer	LCDR Robert T Fischer	RT
	Spouse, Helen	
	LT Steve Kohler	Skunk
	Spouse, Lauren	
	LT Bob Roberts	Butt Chin
	LT Jon Zachery	Stretch
	Spouse, Teresa	
	LTJG Mike Allison	Alice
	Spouse, Amy	
	LTJG Larry Monday	Tuesday
	LTJG Butch Felder	Botch

VA-92 Warhorse Officers 1971 Workups

Commanding Officer[1]	CDR Fuller	
Commanding Officer	CDR Leroy Fant	Little Lord, CO, Skipper
	Spouse, Sarah	
Executive Officer	LCDR Dave Davison	Double D, XO, the exec.
	Spouse, Laura	
Operations Officer	LCDR Simon Toliver	Simp
	Spouse, Maryann	
Maintenance Officer	LCDR Mark Wakefield	EC
	Spouse, Deborah	
Administrative Officer	LT Harvey Engel	Not an Angel, Not
	Spouse, Naomi	
Assistant Maintenance	LT Howie Wisdom	Wiz
	Spouse, Tara	
Flight Schedule Officer	LT Rob White**	Blackey
	Spouse, Carolyn	
Weapons Training	LT Jon Zachery	Stretch
	Spouse, Teresa	

Flight Records Manager	LT Robert Stoll **	Troll
LTJG Mike Allison	Spouse, Amy	Alice
Personnel Officer	LTJG Nat Newsome	Nose
	Spouse, Monica	
Line Division Officer	LTJG Terry Foster	Nooner
	Lydia	
	LTJG Oliver Mason	Skippy
	Spouse, Wanda	Alice
	LTJG Cal Mudd	Mudder
	LTJG Walt Short	Stump
	LTJG Eli Banks	Bee, etc.

[1] Fuller was replaced by Fant.
** LTs Rob White and Robert Stoll were called the Robsey twins.

Navy Rank Abbreviations

Officer

Captain, O-6	CAPT
Commander, O-5	CDR
Lieutenant Commander, O-4	LCDR
Lieutenant, O-3	LT
Lieutenant (Junior Grade), O-2	JG
Ensign, O-1	ENS

Enlisted

Seaman Recruit, E-1	SR
Seaman Apprentice, E-2	SA
Seaman, E-3	SN
Petty Officer Third Class, E-4	PO3
Petty Officer Second Class, E-5	PO2
Petty Officer First Class, E-6	PO1
Chief Petty Officer	CPO

Glossary of Terms

AAA	Antiaircraft artillery, also called triple A and flak.
airdales	On an aircraft carrier, the sailors assigned to the airwing.
airwing	Squadrons assigned to an aircraft carrier grouped and commanded by a senior officer.
Alpha strike	In a low-threat environment, targets could be attacked by flights of two to four aircraft. In a high-threat environment, sixteen to twenty-four bombers would gaggle together while other planes protected the gaggle from MiGs and SAMs.
AOM	All officers meeting. Pilots and ground officers.
APM	All pilots meeting.
bingo	To divert ashore due to a problem with the carrier, with the airplane, or reaching a critical fuel state.

BIT	Built-in test.
BOQ	Bachelor officer's quarters.
CAG	Commander of the airwing.
CMC	Classified material custodian.
CO	Commanding officer, also called the skipper.
COS	Chief of staff.
CPO	Chief petty officer, an E-7.
DMZ	Demilitarized zone.
FAC	Forward air controller, generally in a small propeller-driven plane, whose job was to spot targets on the ground and direct bombers onto the targets.
FENAB	Field Naval Aviator Evaluation Board. Pronounced *fee nab*. A board constituted to evaluate the performance of an aviator after a mishap to determine if the aviator was at fault, and if so, to make a recommendation as to whether or not he should continue on flight status.
FUBAR	Fouled up beyond all recognition.
ground pounder	Maintenance officer, not a pilot.
Ironhand	The anti-SAM mission. Ironhand aircraft equipped with Shrike protected a group of bombers from surface-to-air missiles.

JBD	Jet blast deflector, a slab of steel raised from the flight deck behind an aircraft on the catapult. It is slanted and deflects the blast from a jet engine up and away from planes behind the cat.
JG	Junior grade appended to the navy rank of lieutenant. A lieutenant (JG) wore one silver bar; a full lieutenant wore two.
JO	Junior officer.
JOB	Junior officer bunkroom.
JOPA	Junior Officer Protective Association, a fictious organization incapable of protecting anyone but affording junior officers an opportunity to gripe about the capricious orders issued by senior officers.
KIA	Killed in action.
MIA	Missing in action.
nighter	Night flight.
OCS	Officer candidate school.
OP	Observation post.
ops	Operations.
ops, or ops O	Operations officer.
PO	Petty officer (navy-enlisted rank).
PC	Postal clerk.
QA	Quality assurance.

SAM	Surface-to-air missile.
SDO	Squadron duty officer.
Shrike	An air-to-ground missile carried by US aircraft designed to home on the radars guiding SAMs or AAA fire against US planes.
SOP	Standard operating procedure.
TIAS	Target Identification and Acquisition System.
TS	Top secret.
VC	Viet Cong.
XO	Executive officer.

CPSIA information can be obtained
at www.ICGtesting.com
Printed in the USA
LVHW012029290321
682890LV00002B/139